DESIGN ANALYSIS OF THERMAL SYSTEMS

DESIGN ANALYSIS OF THERMAL SYSTEMS

Robert F. Boehm

University of Utah

John Wiley & Sons

New York · Chichester · Brisbane · Toronto · Singapore

To Marcia, Debbie, Chris, and the Corvairs, who keep my life full and my wallet empty.

Library of Congress Cataloging-in-Publication Data:

Boehm, R. F.

    Design analysis of thermal systems.

    Includes bibliographies and index.
    1. Heat engineering. 2. Engineering
design. I. Title.
TJ260.B57 1987   621.402   86-32609
ISBN 0-471-83204-9

Printed in the United States of America

10 9 8 7 6 5 4 3 2 1

PREFACE

This book is an outgrowth of my teaching of a senior-level thermal systems design course over the last several years at the University of Utah. Generally, the courses in design are difficult for many people to teach, and this one was that way for me. There were two complicating factors. First, I started putting this course together as a fairly green assistant professor. I had had some industrial experience with a few years at the General Electric Atomic Power Equipment Department (later called Nuclear Energy Division) and a number of summer engineering jobs. But it was still a difficult transition from being a doctoral student to being a professor who is supposed to know all things about all things.

Second, there were new currents moving in the field. Computers were proving their value as an engineering tool in everything from characterizing materials to guiding us to the moon. But it seemed that the field of thermal systems was among the last to reap the benefits of this technology. Certainly there were exceptions. Finite difference and, later, finite element techniques were proving to be key tools in many heat transfer analyses. Modeling of fluid flow in a variety of fields was given a considerable boost from a number of breakthroughs during the space race and the associated technological binge the world was experiencing. Even the modeling of control systems seemed to move along rapidly. Power plants, however, were still designed by "hand." Even today, cooling towers continue to be designed by many companies from graphical techniques developed before most of us can remember.

Hence, to aid my own education on thermal systems, and to see where the field might go if computer-aided design techniques were applied, I began studying this field with more than a little interest. It finally got to the point where the file was overflowing with material, and I had to make the decision to do something with it. In 1984-1985 I went on sabbatical leave and began this book.

A general concept of the course and this book developed quite naturally. In this concept, it seemed that the book should draw heavily on the basic analysis courses that students have in the sophomore, junior, and senior years of their engineering curricula. Second, the text should give insights to some of the practical aspects of design that too often are not found in the analysis courses Overall, the course should generally deal with topics in a computer-aided context. Finally, the student should be introduced to the thermal design literature so that the full power of the various techniques can be studied in detail, if desired.

From my course work in thermal systems design, I had ideas about what specific topics I wanted to include in the text. Of course, there should be a discussion of the practical aspects of equipment selection. Of particular concern were some keys to choices between various types of available heat exchangers. For example, too many students leave school without knowing the difference between a floating-head heat exchanger and a U-tube heat exchanger. I addressed questions of this sort with information on selection criteria and costs in a concise format. In this regard, I appreciated Gael Ulrich (of the University of New Hampshire) allowing me to borrow some of his excellent material.

I felt there should be some information on the old, but newly "rediscovered," topic of availability analysis. Workers around the world were showing the value of this concept for the analysis of complex systems where design choices had to be made between energy in various forms. Of course, the cogeneration example is one that illustrates the concept of economic trade-offs between heat and work, and this is used in the text. I think availability will become more of a generally applied tool as we all better understand the concepts and their power in the solution of a wide range of problems.

Flowsheeting is a topic that is given little thought in the systems analysis performed in an introductory thermodynamics course. However, as the system becomes larger and more complicated, the definition of an appropriate flowsheet can be an extremely critical step in the analysis of a system. Many of the questions involved in setting up an appropriate flowsheet have been addressed in the development of powerful codes like ASPEN, and this is an area where research continues. An introduction to flowsheeting ideas is also included.

One topic that is not often covered in basic courses is costs and their effects on the appropriate design selections. Cost data are very difficult to encapsulate accurately, both at a given point in time and in a few pages of a book. The approach used here is the simplified power function form used in the chemical engineering literature. One of the problems with this approach is that the student may view the limited data to be both precise and all inclusive and thus not be aware of the vast differences in costs that can result from various materials of construction, surface finish, or all of the other aspects that can have profound effects on costs. In spite of all of the admitted limitations, the method presented here is chosen as a good "first-cut" approach to the problem of trying to attach costs to components shown on a block diagram. Even the question of which components' prices should be included is a perplexing one. A large number of types of equipment are given in Appendix D, but key items are almost certainly missing. In spite of this, the problem of gathering cost information to perform preliminary design analysis with this information should be greatly simplified compared with starting from "scratch."

Estimates of future costs of energy are likewise very difficult to predict, but this topic is also important to the choices of design options. The data given here are the best estimates of people who make the prediction of prices in their business. Those of us who have watched energy prices through the 1970s and 1980s know how unpredictable these data can be. As the energy cost estimates included here become more dated, users of the book should seek more current information.

A brief chapter on general economic analysis is included. At many schools, these ideas are covered much more thoroughly in an engineering economy course than is the case in this text.

Numerical analysis topics are covered in the text. Included are curve fitting and equation solving in Chapter 4 and optimization techniques in Chapter 9. These chapters share with all of the others the fact that complete texts have been written on subtopics of each chapter. I have attempted to distill the information to give a qualitative feel for approaches represented by the various techniques. Not all techniques are necessarily included. Some that may be very important to various instructors undoubtedly do not appear. Hence, in these topics, as well as in virtually all others, the instructor should supplement the information given here to emphasize important points to him/her.

Curve fits or other correlation information to be used for the prediction of properties are given in Appendix B. At one time or another, we have all curve fit thermophysical properties. In the tabulations included here, I have attempted to make available published data to simplify this thrust.

From the initial stages of the drafting of this book, the plan was to make it concise and qualitative in its approach to the variety of topics covered. However, I wanted to give the student/worker insight to the rich literature spanning both the mechanical and chemical engineering fields. Hence, you will find that every chapter addresses topics that could be covered in complete texts (most, in fact, are the topic of complete books elsewhere), as well as a number of design related papers that may not have found their way into the archival literature. On the other hand, each references section has many more entries than most students will be inclined to consult. Students will find that this book does not have all the information they may want stated explicitly, but they will be able to find it through the use of this text, if they so desire. Some may not want to be bothered by supplementing this text with library materials, but to understand a given topic fully this will usually have to be done.

Several items are not in the text because they have purposefully been left out. Perhaps most obvious is the almost total lack of concepts covered in engineering fundamentals courses. For example, do not look for treatments of the LMTD or Effectiveness-NTU techniques of heat exchanger analysis. They are not here. This is not to imply that there is no need for this information in a design course, but rather it is a recognition that this information is conveniently available elsewhere to students. In fact, I usually review some of the basic material in this course. There are problems in the most chapters that are of a review nature.

In my view, the most important aspect of the student experience in design is the application of the various concepts in a project setting. The book does not carry this aspect through to completion. Only on a teacher-worker level can this work. Decisions may have to be made at each step of the way regarding directions, approximations, desired results, and so on. We may someday find thermodynamics and other courses taught by programmed learning (through computer interaction or special text), but the application of these learning concepts to the design of systems will be among the last to make this transition. There is no replacing the professor or experienced engineer in the day-to-day development of a design. This text addresses topics that should facilitate the process, but these cannot replace the process.

I want to acknowledge people who were helpful in the development of this text, realizing that not all will be mentioned who should be. It all started with John Wiley & Sons' willingness to take on the project, and I want to thank them first. Engineering editors Bill Stenquist and, later, Charity Robey were great facilitators in the overall process. Numerous people at Sandia National Laboratories, Livermore, where I was on sabbatical leave when the drafting began, were willing to take time from their own to look over preliminary notes and make comments. Chuck Hartwig, Jack Swearengen, Joe Iannucci, Jim Dirks, and others are among those in this group. Sandy Baum, an industrial engineering colleague of mine, looked over the engineering economy section. He eliminated several serious errors I had there. I only wish I could capture his enthusiastic expositions on those topics. A number of students, including the brave ones who took the ME 562 course or assisted me when the notes were developing, are especially appreciated. I note particularly Turhan Çoban.

Most importantly, though, I acknowledge my prime proofreader, wife, and friend, Marcia. She gave considerable amounts of both time and encouragement thoughout the whole process. Without her, this book would not have been completed.

My whole career in general and this book in particular have benefited from the guidance and examples set by several advisers and colleagues. Included are Chang-Lin Tien, John Lienhard IV, Frank Kreith, and H. R. (Bob) Jacobs. Thanks, guys.

A final note. Items that probably should have been included in the text were not. Errors are present, but I do not know where they are now. I will appreciate feedback on both aspects of the book.

May your design experiences be among your best.

January 1987   *Robert F. Boehm*

A Note on the Production of this Text

This text, with the exception of Figures 3-10 and 6-1, was produced entirely on an Apple Macintosh personal computer and printed on an Apple Laserwriter. The computer was the 512 k memory machine with dual single-sided disk drives (affectionately referred to by some as the "hummer"). Software applications used were MacWrite for word processing, MacDraw for rendering the line diagrams, and Microsoft Chart for the plotting done in Chapter 6. In all cases the application Switcher was used to go between other applications and the word processer. Printing was done entirely in Geneva 9, 10, and 12 point fonts with laser font substitution. Final rendering was done on 8 1/2 by 14-in. sheets, which were used for reproduction at Wiley. Appreciation is expressed to the University of Utah for making this equipment available.

About the Author

Robert F. Boehm was born in Portland, Oregon in 1940 and was raised in the state of Washington. He attended Washington State University, receiving a B.S.M.E. degree in 1962 and an M.S.M.E. degree in 1964. He then joined the General Electric Company, Atomic Power Equipment Department. He left GE to pursue a Ph.D. at the University of California at Berkeley where he received the degree in 1968. He then accepted a position at the University of Utah where he is now Professor of Mechanical Engineering. During his tenure at the University of Utah he has served as chairman of the Mechanical and Industrial Engineering Department. The 1984-1985 academic year was spent at Sandia National Laboratories, Livermore, California, where this text was started. He is the author or coauthor of nearly 100 technical articles, 2 other texts, and approximately 10 chapters in texts on heat transfer and thermal systems. Research interests include experimental and numerical heat transfer studies and analysis of thermal systems, with particular emphasis in applications to energy conversion. He is a Fellow of the American Society of Mechanical Engineers and serves as Technical Editor of the ASME's *Journal of Solar Energy Engineering.* Dr. Boehm is a registered professional engineer in the state of California.

CONTENTS

Contents

9 INTRODUCTION TO OPTIMIZATION TECHNIQUES   187

APPENDIX A: DESIGN PROJECT SUGGESTIONS   217

CHAPTER 1

THE DESIGN ANALYSIS PROCESS

1.1 WHAT IS DESIGN?

Much has been written about the topic of design. Design has always been important in all engineering practice, but its relative importance as a distinct field in engineering curricula has ebbed and flowed over the years. Generally, the design process involves the application of concepts from engineering science topics in a generally specified manner coupled with a creative touch.

Successful design is a collection of several processes. First, insight into the desired end result is necessary. This step might be called *conception.* Second, the ways in which that end result might be accomplished must be defined. The term *synthesis* might be applied to this step. Finally, a significant amount of *analysis* is often needed to supplement the first two steps to bring the design to reality. Analysis can also supplement the synthesis function.

The insight noted in the first step depends very strongly on a hard-to-define characteristic called *creativity.* Does a person have creativity at birth, or is it something that can be learned? Since there are numerous successful inventors (invention can be a form of design) who do not have a formal education, it would seem that some people are born with a natural ability in this area. There are varying opinions about whether or not it can be learned. The second step, called synthesis, is obviously very important. It requires both learned information and creative insight. A person can come a long way in setting up a more efficient thermodynamic system by studying the various kinds of processes possible and the factors that influence their efficiency. From time to time, though, creative insight will render a clear breakthrough in a given design. Finally, the third step related to analysis is clearly something that can be learned. The analysis function can find application in the synthesis step, causing a gray area in the definition of these two terms.

Compare distinctions between mechanical design and thermal process design. Like many, this division may be somewhat arbitrary, but it is important to the thrust of this text. Some elaboration on this comparison as well as the elements of the creative aspect in each category are in order.

Suppose that a firm wants to manufacture and sell a new type of mechanical can opener. Suppose further that in order to have a market edge, the new device will have an operational advantage over existing designs and will sell for a competitive price (this may actually be a higher or lower price than the other openers on the market, depending upon the perceived operational advantage of the new design). Creative work must then occur in two areas. First, the operational advantage must be conceived and reduced to practice. Some people think that the creative procedures involved in this step are not, in general, easily categorized or learned. Once a device is designed, the processes used in the manufacturing steps must be determined. Although there are avenues to demonstrate creative genius in these aspects, several dimensions can be developed into a specific technology and can be categorized and learned. If the product is truly successful, hundreds of thousands might be manufactured and sold, so the clever manufacturing of the device may be just as important as its original invention.

In the example of the can opener, as well as other more involved mechanical devices, creative aspects can be the most critical in the device conception stage and can be of somewhat lesser importance in the other facets of bringing the product to the consumer. The relative importance of the upfront creativity is lessened in more complicated mechanical systems and thermal systems where refinement of designs may make up a large portion of the creative process. In these examples, the modeling of the devices or systems may be extremely valuable, enabling the overall improvement of function and performance.

Computer-aided engineering can be of great value in the solution of mechanical design projects. Often this involves a graphical representation of the part or device via the computer. See the left-hand side of Figure 1.1. This can ultimately be used for assisting with the manufacture of the device. In addition, as the design and analysis functions take place, the graphical representations can be used for stress, electric/magnetic field, and/or temperature analyses of the device.

On the other hand, the simulation of thermal systems often involves the synthesis of components into an overall system. See the right-hand side of Figure 1.1. This is a subset of the overall category of process design. In contrast to mechanical design, the computer can be used here to simulate processes. Both fields can share a need for optimization and other types of numerical analysis, but there is a fundamental difference between the two.

Consider applications to the design of thermal devices and systems. A system may be very large and have a single application. An example of this is a giant mine-mouth power plant in Wyoming. Alternatively, it could be some system produced in large numbers, such as a new refrigeration unit to be applied as an automobile air conditioner.

What are some points of contrast between most thermal systems and the simple mechanical devices as illustrated above with the can opener example? In contrasting them, there is always the fear of overgeneralizing, but certain aspects are more often true than not.

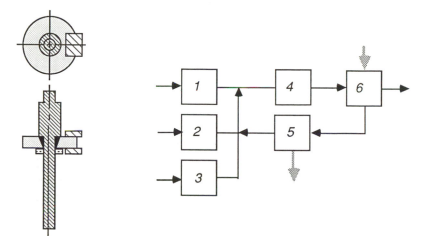

Figure 1.1 Computer-aided engineering can play a vital part in design. In mechanical design this usually involves a graphical representation of the part or assembly (left). Process representations in the computer are important in thermal system design (right).

1. Thermal systems generally involve components that can be categorized. For example, pumps, heat exchangers, turbines, mixers, and other items are found in a large number of thermal systems.

2. In fact, thermal systems often involve a large number of components in one design.

3. The presence of a number of components means that there is a large number (perhaps in the thousands) of parameters that must be set, either arbitrarily or in relation to other aspects of the design.

4. Parameters to be set in the design usually affect both capital and operating costs, the latter being greatly influenced by, and often totally dominated by, energy costs.

5. The general system that is to be designed may be well defined, but it can still involve a major design activity, possibly including modeling, to satisfy the project objectives.

6. Modeling of thermal systems can be systematized to a large degree.

The topic of thermal systems design owes much to W. Stoecker of the University of Illinois. His initial treatise (Stoecker, 1971) on the topic was a pioneering work in this area.

1.2 CATEGORIES OF DESIGN

1.2.1 Empirical versus Analytical

Thermal systems have been designed for hundreds of years. Their development has been due, by and large, to *empirical techniques*. Perhaps a blacksmith fashioned a firebox to a size that he felt would be satisfactory for a given application. Or perhaps an engineer designed a boiler using available tubing lengths and sheet metal sizes without much thought about what else might be possible. The automobile engine is an excellent example of how a fairly complex device can be developed almost entirely from empirical means.

During the last half century, empirical means are gradually being replaced by *analytical techniques*. This trend was aided in its initial stages by the development of analytical methods in heat transfer, fluid mechanics, chemical processes, and thermodynamics, which, in turn, evolved from empiricisms themselves. The analytical trend was given a dramatic boost in the 1950s with the development of the high-speed digital computer. More recently, the application of optimization techniques has proved to be extremely valuable.

1.2.2 Nonfunctional, Functional, Satisfactory, and Optimal Designs

Regardless of the techniques used for defining the design, the synthesis process will result in either a *functional* or a *nonfunctional* design. If a solar-driven refrigerator is designed, but it will not actually cool, then the design is said to be *nonfunctional.* However, if the refrigerator does perform in some manner as it should (i.e., it cools), the design is termed *functional.* Usually this is not sufficient. While any design is functional or nonfunctional, this distinction is usually obvious. The engineer is concerned with a great deal more than simply a functional design.

A *satisfactory* design is drawn from the class of functional designs. The *satisfactory* design (there may actually be several in specific situations) meets some kind of stipulated criteria. Usually the stipulation is on some aspect of performance ("it must cool 6 m^3/s of air from 30°C to 4°C" or "it must produce 4000 kW_e"). This can give rise to a number of designs that all meet the objective. Many times there could be cost factors that dictate a satisfactory design ("it must cost less than $500,000") or, as in the case of the space program, reliability factors, with cost factors

4

seemingly taking a secondary role. (Actually, the manned space program implicitly put a very high value on human life, a stipulation that led to requirements for high reliability.)

The final, and often most desirable, category is that of the *optimal* design. Here some very specific restrictions may be placed upon the final product, and usually these are in the form of some stipulation on cost. For example, a power unit that produces 4000 kWe may be desired, but an optimal design that minimizes life cycle costs over an anticipated 20-year life is also desired. There may be virtually thousands of satisfactory designs (ones that produce 4000 kWe) but only one (or at most a few) optimal designs.

The categories of designs can be shown on a solution space. See Figure 1.2 for an example involving two independent design variables.

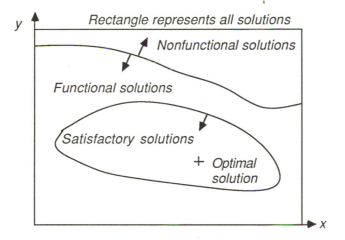

Figure 1.2 A hypothetical example showing various system solutions.

Too often, engineering practice is satisfied with simply a satisfactory design (and sometimes just a functional one!) for a thermal system. This has been due to a number of factors, including low energy costs and the lack of design analysis tools.

The impact of energy costs is easily envisioned. If operating costs are negligible, then first (capital) costs become the determining factor. In this case, minimization of capital costs may minimize life cycle costs. However, as energy and other operational costs rise, designs that simply minimize first costs fall into the category of functional or, at best, satisfactory design possibilities.

Design analysis tools have not been present historically. This has changed with the advent of the computer. Included are both hardware and software impacts. Both of these aspects have benefited from considerable development over the last twenty years. Desktop computers are now more powerful than were the largest of all the supercomputers less than fifteen years ago. Software libraries become more extensive with each passing year. The coupling of the engineering science knowledge with the existing computer software can make a powerful combination. It is to this end that the present text is dedicated.

1.3 ELEMENTS IN THE DESIGN ANALYSIS OF THERMAL SYSTEMS

There are several steps in the design process for thermal systems. In the discussion that follows, a system is assumed to be made up of more than one process, so that process elements are the foundation of the synthesis procedure. As shown in Figure 1.3, one way of looking at the general approach is to assume that it is made of three general steps.

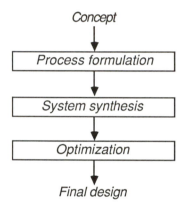

Figure 1.3 A broad categorization of the three basic elements in the definition of an optimal design of a thermal system: process formulation, system synthesis, and optimization.

First, and by far the most important aspect, is the necessity of having a good understanding of the physics of the various processes and other elements that will be needed to make up the system. Without an understanding of the basic ideas of heat transfer, thermodynamics, chemical processes, and fluid mechanics, for example, it would be improbable that a person would be very successful in designing a thermal system. (More will be said about this point later in this chapter.) This will be denoted here as *process formulation.*

Second, this knowledge of physics, including a great deal of supplementary information of a more empirical nature, is also needed to build the elements into the kind of system necessary to accomplish the desired result. For our purposes, this will be termed *system synthesis.*

Finally, the system model should be exercised to yield the optimal, or at least a nearly optimal, solution. This stage is defined here as the *optimization* block.

For the most part, it is assumed that someone interested in the design of thermal systems has already been thoroughly introduced to the physics of process formulation and, to some degree, system synthesis. For example, consider the system design of a regenerated Brayton engine with a bottoming steam cycle. It should be obvious that some knowledge is necessary of what components constitute these kinds of cycles. For this, it is assumed that knowledge of the LMTD and Effectiveness-NTU methods for sizing of heat exchangers is reasonably understood also. Insight to the general performance prediction theory of turbines, compressors, and pumps is also necessary.

The techniques related to the system design aspects can then be used to formulate the optimal system. Stress will always be placed on the optimal design. Although some systems may not have an optimal form, this occurs very infrequently. Usually the problem is not that there is not an optimal (or at least a "nearly optimal") solution. Instead, it may be that the optimum condition is difficult to define fully and to find. The first of these problems is often due to imperfect understanding of the physical problem, whereas the second may be due to the mathematical complexities of a solution technique.

A more accurate representation of the analysis needed to arrive at a desired design is shown in Figure 1.4. The *design concept* consists of defining what is to be done by the design and how it might be accomplished. Not infrequently, there may be several approaches that may satisfy a given input-to-output requirement. After analysis, one may be shown to be preferable, but usually this insight is not known in advance.

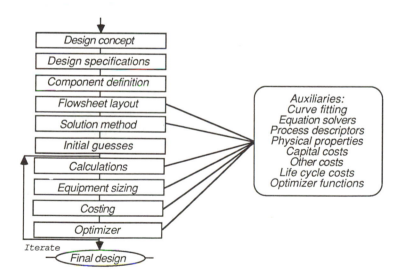

Figure 1.4 The analysis performed in pursuit of a final design of a thermal system can involve a large number of steps. In this diagram, each block can represent many steps that are fully defined here, but most of these are described in detail later in this text.

The *design specifications* step, in which the output is explicitly defined, is especially critical. It is hard to achieve a goal that is not well specified. Not only is it important to note what is desired, but it is also necessary to specify what is not wanted. A fin design for a tube that maximizes heat transfer may also be the largest one that can be built. Some concern about the size is probably also of value. Often, consideration of component costs in the overall analysis will help control many of these types of factors.

Component definition requires the application of concepts of the physics of the various aspects of the overall system. The usual goal is to determine the necessary mathematical relation(s) in order to define fully each process component. While it is desired that this be done in closed form (i.e., based upon theoretical concepts) so that the model will cover the range of operating parameters, in fact, it is often necessary to apply empirical relations. Even in designs using heat exchangers, where the performance of the latter is fairly well understood, the pressure drop and heat transfer coefficient characteristics will undoubtedly be represented empirically.

The *flowsheet layout* step is important for those design problems where there is a large number of process components. If a system has only one component, or at most only a very few, the formality represented by this block on the diagram is not so critical. For complex plants with a large number of interconnected components, this step is absolutely necessary to allow analysis to proceed. Even in simple systems involving only a small number of processes, it is desirable to lay out the flow diagram to aid in the understanding of system-operating characteristics.

A large number of variations is represented by the block denoted *solution method.* As with the flowsheet layout step, this block can represent a very complex set of procedures when the system involves a large number of processes. Typically, these procedures will involve a step-by-step approach to cast the block diagram into a form that will yield a mathematical solution.

At this point in the scheme laid out in Figure 1.4, the system has been defined, the mathematical functions for all of the processes have been found, and the general solution technique has been laid out. Now a need will often exist for *initial guesses* for a large number of the problem variables so that the solution can proceed. Obviously, some systems can be analyzed directly from the given inputs, and this step is not necessary in these cases. However, in the more complicated systems, this step can be very important. In some instances, the ultimate success of the solution is dependent upon reasonable choices of initial guesses.

The step denoted *calculations* is where the computer takes over. Generally, what is represented here is a system model where variable inputs, possibly including component sizes, some process inlet pressures, temperatures, and so on, are then used in the calculation procedure to find all of the system interactions. This latter category may include power inputs and outputs, heat additions and rejections, and all other inlet and all outlet mass flow stream definitions.

In many analyses, the solution will be carried out on a unit-mass basis. Certainly in these situations, but also in a number of other cases, there is a point where the size of the components must be defined. How many kilowatts must be produced by that generator? What is the total duty of this heat exchanger? These kinds of questions must be answered at this point if they have not been addressed earlier in the analysis. This function is denoted on the diagram by the block entitled *equipment sizing*.

Costing is the step where a dollar value is attached to each of the aspects of the plant. Usually, this not only will involve the assigning of the costs to the various pieces of equipment, but it also should include the estimation of the operating costs and all other factors that will give the "bottom line" on the overall price of the plant. This step is always important; and it can be extremely involved, particularly in the final design process. In this text, only the first step ("preliminary cost estimation") is emphasized. Capital costs are taken as the appropriate economic indicator in this approach. While this simplification can yield misleading results in some situations, this is generally a good first-step analysis for most thermal systems. Operating and maintenance costs, where significantly different for various design options, can often be considered in another complete evaluation of the designs.

At this point on the diagram in Figure 1.4, a single system has been defined and all of the costs evaluated. This system could be quite far from the optimum form. Moving the solution toward the desired overall result is the function of the *optimizer*.

1.4 THE APPROACH OUTLINED IN THIS TEXT

1.4.1 Summary of Topics Covered Here

In what follows, an attempt is made to summarize much of the material that is needed to model thermal devices and systems. It is assumed that the reader has a background in mathematics, fluid mechanics, thermodynamics, heat transfer, chemical processes, engineering economics, and computer programming even if this background is very minimal. There are too many items that need to be discussed about other aspects of steady-state system simulations to spend much time here addressing these fundamental topics. The coverage of this book and how it is assumed to fit with more familiar topics is depicted in Figure 1.5.

Chapter 2 covers aspects of selecting pumps, fans, compressors, and other fluid flow equipment. While the theoretical aspects of fluid mechanics as well as the thermodynamic analysis of devices like compressors or turbines would have been introduced to the engineer or student at this point, too often, practical distinctions between generic types of equipment are not covered in introductory courses. For example, what are the basic types of compressors; and when should one type be used rather than another type? In many instances, "rules of thumb" are given to help determine which general type of equipment will be most appropriate.

These distinctions may impact some aspect of the physical analysis (e.g., one type of device may have a higher efficiency than another), and these distinctions almost always have important cost considerations. Estimating the cost of capital equipment is treated in Chapter 6, but a key to the application of cost data is through the appropriate equipment selection criteria. Example problems and problems at the end of the chapter can be used to review basic methods of analysis covered in previous courses.

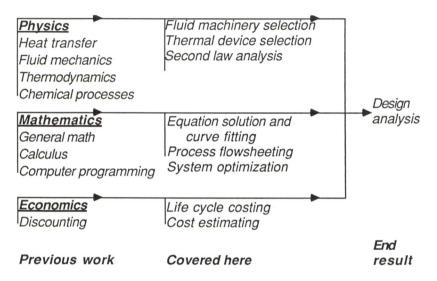

Figure 1.5 The topics of this book (shown in the second column here) are assumed to supplement previous knowledge to enable the reader to perform a design analysis.

Selection of heat exchange devices is the focus of Chapter 3. Of course, this includes a general description of the various types of heat exchangers that are commercially available, but also covered are combinations of equipment items that are combined into subsystems, like those used for heat rejection functions to the atmosphere. Wherever possible, rules of thumb are given to give insight into when one type would be used compared to another. While generalities of cost comparisons are listed here, more specific costing information is given in Chapter 6. As is true for fluid flow devices in Chapter 2, example problems and problems at the end of the chapter can be used to review the methods of analysis of heat exchange devices.

In Chapter 4, some important numerical analysis tools used often in thermal system simulation are discussed. These include means of root finding, curve fitting, interpolation, and solution of systems of equations. As with virtually all of the chapters given in this book, the coverage of this chapter could be the topics found in a complete text. The emphasis here is on developing physical insights for application of a few techniques. A coverage of the mathematical basis for the techniques, their absolute accuracy bounds, and alternative approaches will have to be sought out in other literature. A key to additional information is given at the end of the chapter.

Most design decisions are made on the basis of economics. While a variety of techniques can be used for determining project costs, the emphasis here is on simplified approaches. The ones addressed can be used easily and applied widely to allow cost comparison between various options. Topics that are often found to be included under the title "Engineering Economics" are given in Chapter 5. Some simplified techniques that can be used for determining the value of systems are given, as these sometimes can give quick answers to questions related to investments. Drawbacks to these shortcut approaches are also outlined. Most of the emphasis in this chapter is given to the method of life cycle cost analysis. Some comparisons to other techniques used to evaluate project overall economic value are given also.

A very important part of performing an economic analysis of a project is the cost of the major pieces of equipment used. This step in the design analysis process might be called "Preliminary Cost Estimation," and it is covered in Chapter 6. A key to the technique is a tabulation of pertinent cost data for a large number of equipment pieces in a variety of sizes. In Appendix D, an extensive table of this kind of information is given. Costs can become outdated quite quickly, and the techniques for updating most historical data are also given in Chapter 6. References to more detailed and accurate sources of costing information are given so that "second cut" designs can be evaluated if desired. Also included in this chapter are some professionally performed predictions of energy costs for the next couple of decades. This is another key ingredient to the determination of the overall economic viability of a project.

In Chapter 7, some concepts from applied thermodynamics are presented. The focus here is on Second Law applications. There is a growing emphasis on the use of the concept of thermodynamic availability in the analysis of thermal systems. Although many readers may have covered this topic before, the design implications are stressed here. The value of this concept in making comparisons between work and heat in systems, as well as evaluating maximum performance possibilities, is emphasized.

As shown in Figure 1.4, generally the next step in the building of a simulation model is the representation of a series of processes into a mathematical description of the whole system. This has to be set up in a way that analysis can follow to determine the performance of the system. Chapter 8, "Process Flowsheeting," addresses some of the ways that this can be done. Usually, the resulting system is one of two types: one where the calculations can be performed in a sequential manner from the beginning of the system to the end; or one where the calculations must be done in a simultaneous manner, top to bottom. While the general design process usually requires the latter approach, many times the sequential technique can be used in an iterative way to determine the same end result.

11

The latter approach is often much easier to set up for solution.

A final major topic of this text is an introduction to optimization methods. This is covered in Chapter 9. As is the case with each of the chapters, the topic touched here fills volumes of other texts. Selected aspects are included that illustrate the topic in a conceptual manner. References are given to allow the reader to pursue other approaches of a different type or of more complexity.

Additional material is included in the appendices of this book. A list of project possibilities for applying the concepts given in this text is included as Appendix A. Some information on correlations that can be used for generating computer routines that determine thermodynamic and thermophysical properties is given in Appendix B. References to sources of computer routines for property evaluation are included also. Appendix C gives a catalog of curve fitting functions to aid in the selection of appropriate forms for specific sets of data. This material is supplement to the information given in Chapter 4. As already mentioned, Appendix D gives capital-cost data to be used with the ideas outlined in Chapter 6. Finally, a list of a few (of literally thousands) government documents that give data or show techniques pertinent to the design of thermal systems is given in Appendix E.

1.4.2 Steady-State Analysis

One important aspect should be noted before ending this introductory chapter. <u>The emphasis of this text is on steady-state simulations</u>. While the point can certainly be made that virtually all systems have some element of transient behavior and are constantly cycling between off-design states in actual operation, the importance of mastering steady-state simulations is fundamental. In spite of the presence of the generally transient behavior of systems, the fact remains that practical design is based on steady-state performance.

Much of what is discussed here is pertinent to the transient analysis of systems. Often it is sufficient to equate a steady-state balance of flow quantities (specie, mass, energy, or availability) to a time rate-of-change of the pertinent system state quantity (specie, mass, energy, or availability). First-order, ordinary-differential-equation systems often result. A number of texts treat dynamic system analysis, and the reader who is seeking information on the dynamic analysis should have no trouble finding a great deal of background material on this topic.

1.4.3 Conclusion

In all, this text should go a long way in building the reader's experiential base on thermal systems simulation. As in most fields, in the design of thermal systems, there is no substitute for understanding the analytical tools available and gaining experience in applying those tools. Now is the time to start!

REFERENCE

Stoecker, W., 1971, DESIGN OF THERMAL SYSTEMS, McGraw-Hill Book
Company, New York.

PROBLEMS

1.1 Consider a simple, ideal Rankine cycle operating between pressures of 0.1 and 50 atm. 100°C of superheat is present entering the turbine (i.e., if the saturation temperature is T_{sat}; then this temperature is $100 + T_{sat}$). It is desired to add a single stage of closed regeneration to the cycle. Suppose the pressure at the extraction point is taken to be 20 atm.

(a) Is this a functional or a nonfunctional regeneration design?

(b) What is the thermal efficiency of this configuration?

(c) If the optimal design is taken to be the one that demonstrates maximum cycle efficiency, find the appropriate extraction pressure.

(d) What is the cycle efficiency of the optimal design?

1.2 It is desired to heat 50,000 lbm/hr of a motor oil from 70 to 180°F using steam condensing at 1 atm pressure. 3/4-in.-ID, 13/16-in.-OD copper tubes 8 ft long are readily available.

(a) Specify a satisfactory design for a heat exchanger that would accomplish this heating requirement.

(b) In practice, what stipulation(s) might be used to define an optimal design? Speculate how you might accomplish this design. (The optimal design is not required.)

1.3 A 400-ft-long pipe is to be run from a receiver tank on a water pump to a valve that exhausts to atmospheric pressure. No more than a 2-psi pressure drop is to be allowed in the line between the pump and the valve. See Figure 1.6.

(a) Plot the maximum flow rate possible against the pipe diameter.

If a 200-gpm flow is desired, show on your plot the following:

(b) the functional design(s);

(c) the nonfunctional design(s);

13

(d) the satisfactory design(s); and

(e) the optimal design(s).

(f) If one, or more, of the above categories does not have an answer you can show, speculate what kinds of information might be given to enable you to find the answer.

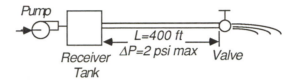

Figure 1.6 Sketch of system discussed in Problem 1.3.

1.4 In a system related to that noted in Problem 1.3, water is pumped through 400 ft of one diameter of pipe to a point where 200 gpm exits through a valve. The pipeline, now constructed of a different diameter, extends from that point another 400 ft to a second valve where a second 200 gpm is to exit at the end of the line. See Figure 1.7. If a total maximum pressure drop of 10 psi from the pump to the second valve can be tolerated, specify some possible diameters for:

(a) the functional design(s); and

(b) the satisfactory design(s).

(c) Assuming that the cost of the pipeline is directly proportional to the diameter of the pipe used and that it is desired to have the least cost pipeline, specify the optimal solution (i.e., give the appropriate pipe diameters to yield a 10-psi pressure drop).

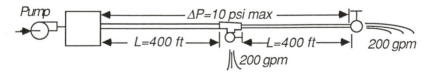

Figure 1.7 Diagram for system described in Problem 1.4.

1.5 Double-glazed windows are a common product used to decrease heat loss through fenestrations. A particular manufacturer's products have 1/4-in. spacing between panes. Neglecting heat loss through the frames, is this a good design assuming preference is given to minimizing heat loss? Support your conclusion with calculations.

14

1.6 Consider a horizontal, stainless steel pipe line of length L. The pipe inside and outside pipe diameters are D_i and D_o, respectively. Assuming steam is condensing in the pipe at temperature T_{sat} due to heat loss to the ambient at temperature T_a, discuss the following points.

    (a) What calculations would be needed to determine a functional solution for decreasing the heat loss from the pipe? (Discuss and illustrate with equations as appropriate.)

    (b) What calculations would be needed to determine a satisfactory solution for decreasing the heat loss from the pipe? [Again discuss and illustrate with equations as appropriate. Specifically note items that are different in this part compared with part **(a)**.]

    (c) Give a reasonable statement of the optimal solution for decreasing of heat loss from the pipe. Indicate what additional information and calculations would be necessary to carry out this portion of the solution.

1.7 A developer wishes to break the speed record for steam-powered automobiles.

    (a) Define several of the general considerations you should stipulate in the design to yield a satisfactory solution. This should include not only the end result(s) desired but also the technical requirements for accomplishing the end result(s).

    (b) Is there an optimal solution here? What factors might be involved in defining a solution of this type?

1.8 Consider combustion of methane, CH_4, with air, taking place at standard atmospheric pressure in a chamber. Define the amount of air required for each of the following situations. In each case, assume that it is desired to achieve temperature leaving the chamber that is greater than 90% of the adiabatic flame temperature (in absolute degrees).

    (a) The functional solution(s).

    (b) The satisfactory solution(s).

    (c) The optimal solution(s).

CHAPTER 2

SELECTION OF
FLUID FLOW EQUIPMENT

2.1 INTRODUCTION

The beginning designer can be faced with a very confusing array of equipment choices in formulating a system design. This problem may be lessened by the engineering practices in many companies. In these practices, the novice is assigned to a very small portion of a well-established design line and is asked to focus on small modifications of a particular kind. The inexperienced person thus actually serves a type of an apprenticeship. This approach tends to ensure that the modified overall design will not function too differently from the original. This is good in that the old design presumably functioned to some degree in a desirable manner. However, aspects of the device or system may be able to function much better if the whole is given a critical examination overall.

Sources of practical information about choosing equipment and the corresponding standard engineering practice in industry are, too often, not in the hands of the newly graduated engineer. Textbooks usually minimize this type of information in favor of a focus on theoretical topics. Other than whatever information may have been gathered in the given industrial organization, perhaps both in the senior engineers' experience and in formal company documents, the easily accessible repositories for this important knowledge are often difficult to find. More typically, available specialized texts contain this type of information. Most frequently, however, the information will surface in trade publications within a given field. Thus, any engineer may find the frequent reading of these kinds of publications to be of great value, and these publications can be of even more value to the novice.

A magazine such as Heating, Ventilating, and Air Conditioning can be of great value to people in the HVAC industry, while a magazine like Power often has articles of great practical interest to engineers in the electrical generation industry. Chemical Engineering is a similar type of publication that focuses on the process industry. Numerous other publications, including the magazines put out by the various technical societies like the American Institute of Chemical Engineers (AIChE), American Society of Mechanical Engineers (ASME), and American Society of Heating,

Refrigerating, and Air Conditioning Engineers (ASHRAE) are available. (Also, each of the technical societies usually publishes one or more archival journals. Examples include the ASME's <u>Journal of Solar Energy Engineering</u> and the <u>AIChE Journal</u>. These publications tend to focus on more fundamental information, but they often contain practical design information for the beginning engineer also.) While each of these publications has a very distinct focus, engineers who design thermal systems in any industry may find much of value in all of these types of publications.

To illustrate in a small way the kinds of information found in these publications, some insights about the selection of equipment are extracted from a variety of typical publications and are summarized here and in Chapter 3.

Some considerations that go into the correct choice of fluid flow equipment are discussed in this chapter. Included here are factors that should be contemplated when specifying pumps, fans, compressors, turbines and other expanders, storage, valves, and piping. Proper choices between types of devices in the first five of these categories are critical in preliminary design studies of systems that handle fluids. On the other hand, a need for any kind of detailed specification of valves and piping is normally not present at the preliminary design stage. Information on these latter two categories of devices is included to provide background for people who are unfamiliar with these topics.

As with virtually all sections of this book, the background material on the topics touched upon fills volumes. Hence, the descriptions given here should be anticipated to be quite abbreviated. References to more information are given at the end.

2.2 PUMPS

The number of different types of pumps available is very large. Each one, in turn, has a number of applications that it can perform. Often, there may be an optimal type of pump for a given situation. At other times, several cost-effective solutions may exist. For the beginning system designer, an introduction to the general categories is critically needed.

Traditionally, pumps have been classified into three general groups: *reciprocating, centrifugal,* and *rotary.* Over the years, a large number of new developments have taken place, and these distinctions have been blurred. As a result, some have chosen to define two general categories of pumps: *kinetic energy forms* and devices that operate on *positive displacement* concepts.

In what follows, pumps will be described in terms of the categories of *centrifugal* and *positive displacement.* Qualitative explanations of performance and costs will be given. Many aspects relating to the full description of the pumps, like actual performance curves, drive options, and materials of construction, will not be included in this discussion.

Much more complete descriptions of pump distinctions and performance can be found in the literature. See, for example, the special report given in <u>Power</u> magazine (O'Keefe, 1972). This excellent summary gives examples of many of the pumps in existence today.

To illustrate a few of the differences between reciprocating, centrifugal, and rotary pumps, consider Figure 2.1. Here the general concepts of operation are shown. In the rotating form, elements continually contain volumes of fluid and physically "push" those volumes through the outlet. Thus, the rotary form is similar to the reciprocating type of pump. The latter type uses the back-and-forth movement of an element to push the pumped fluid through the outlet valve. An inlet valve opens at appropriate points during the movement of the back-and-forth element to allow lower pressure fluid to flow into the chamber. Both the rotary and reciprocating pumps are positive displacement devices. Finally, the centrifugal type of pump accomplishes its pressure boost by imparting kinetic energy to the fluid. In the situation shown in Figure 2.1, the lower pressure fluid enters the pump assembly from the center of the rotor. It flows in a generally radial direction, picking up velocity. As the fluid exits the pump, a pressure rise due to centrifugal force is given. The characteristics of the pump types just described could easily lead to the definition of two categories of pumps: *positive displacement* (both the *rotary* and *reciprocating* devices described here) and the *centrifugal.*

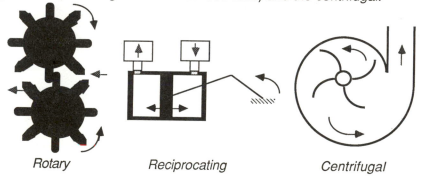

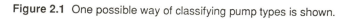

Rotary   Reciprocating   Centrifugal

Figure 2.1 One possible way of classifying pump types is shown.

Although there are many differences between pumps of various types, including the materials of construction, which can have a significant effect on the pump cost, the more important ones for the person performing preliminary designs involve the pressure head range possible from the pump, the fluid flow capacity range of the pump, and the functional relationship between these two variables. Each of the thousands of types of pumps that are available has different kinds of characteristics, so it is impossible to give any specific summary here. However, keep in mind the general differences between the positive displacement and the centrifugal types. Although exceptions can certainly be found, some typical variations are shown in Figure 2.2. Positive displacement types of pumps usually show a very steep characteristic head-versus-capacity curve, while centrifugal pumps usually demonstrate a flatter curve. One exception is the *diaphragm pump* type, which operates on a positive displacement principle. In this device, a flexible diaphragm is moved back and forth, either by some mechanically reciprocating drive or by the admission and venting of a gas. This type of pump can have an output curve more similar to the centrifugal than to the positive displacement types.

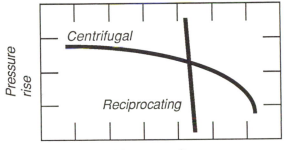

Figure 2.2 A generalization of the pressure-head-vs.-flow-capacity characteristics is shown for pumps of the centrifugal and reciprocating types. The latter tend to be much steeper, although exceptions can be found for both the variations shown here.

Gross simplications are normally made when comparing the performance of the two types of pumps. (i) If <u>high heads</u> are needed, usually a <u>reciprocating</u> pump will be necessary. (ii) If <u>high flows</u> are needed, usually the design will require a <u>centrifugal-type</u> pump. Centrifugal pumps can normally be valved-out (flow cut to zero), but such is not considered good practice with a reciprocating type. In the applications of the latter, a bypass is normally incorporated if zero flows are anticipated through the piping circuit.

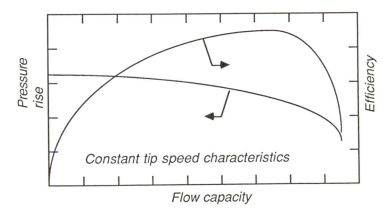

Figure 2.3 A typical pressure-head-vs.-flow-capacity curve for a centrifugal pump operating at constant tip speed. The efficiency characteristics of the pump are superimposed on the pressure head curve.

19

Centrifugal pumps find the greatest number of applications in most plant designs. For this reason, it is of value to indicate typical characteristics for these devices. Figure 2.3 is a qualitative plot of centrifugal pump head characteristics. In addition, another critical piece of information is shown--the *pump efficiency.* This is obviously important in determining the power required to perform a given pumping process.

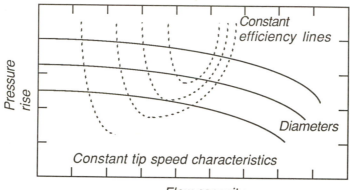

Figure 2.4 A schematic of the performance curves for a family of centrifugal pumps. The solid lines denote the pump head characteristics with flow at various impeller diameters. Lines that are dashed show the map of constant efficiency. All lines on the curve are assumed to be at the same impeller tip speed.

Figure 2.4 shows related curves to those in Figure 2.3. In Figure 2.4, the performance for several impeller diameters is shown. Dashed lines of constant pump efficiency are shown superimposed. Information of this type is highly desirable in any simulation of a system requiring detailed information on the performance of a centrifugal pump.

General comparisons of pumps may be of value beyond the high-flow, low-head (centrifugal) and vice versa (positive displacement) observation made earlier. Typically, the centrifugal is favored where initial cost is very important, or where fluids carrying solids that can erode internal valves are pumped. The centrifugal also demonstrates a fairly uniform output pressure, while a positive displacement pump may give a fluctuating output pressure in tempo with the cadence of the rotors, piston, or whatever the internal mechanism.

Another important characteristic is the *net positive suction head (NPSH)* required for a given pump and the available NPSH in the piping circuit. Although it is critical that all actual installations operate with the available amount of NPSH greater than that needed, this aspect is usually not of critical concern in the preliminary design stage. Note: one often encounters "condensate" or "hot-water" pumps. Normally, these are pumps that have a very low NPSH requirement.

Another type of pump used for special service is the *jet pump.* This device, and its close relative the *ejector,* which is used for moving gases, works without moving parts. See Figure 2.5. A high-pressure fluid is used to move a lower pressure fluid by injecting the former through a combined nozzle/converging-diverging assembly. Note that the high pressure of the propelling stream must be generated by some source, and this will undoubtedly require a separate pump of another type. The jet pump can be used to move a low-pressure fluid, which may be very difficult to handle. Devices of this sort have also been used in locations where high reliability is needed. One example of the latter is an application for moving the core coolant totally within the containment vessel of a nuclear reactor.

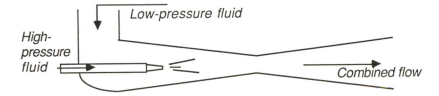

Figure 2.5 Schematic diagram of a jet pump. An ejector is a similar device used for moving gases or vapors.

When designing systems where fluid movement must be accomplished, devices that might not be thought of as "pumps" may have application. Included here is the use of *gas "lifts"* where a gas simultaneously flows through a vertical column of liquid and moves the liquid. Another approach is to *pressurize a tank* holding the liquid with an overlaying gas.

A summary of selection criteria for pumps is given in Table 2.1. This table was abridged from Ulrich (1984). Information is given there not only on the flow/pressure rise ranges, but also on applicable temperature and viscosity ranges. See the original source for similar information on a greater number of pump types and additional criteria on materials of construction.

Further information on the description and applications of pumps can be found in a number of sources. (See, e.g., Hicks and Edwards, 1971; O'Keefe, 1972; Walker, 1972; Neerken, 1974; Karassik et al., 1976; Peters and Timmerhaus, 1980; Pollak, 1980; Stewart and Philbin, 1984; Ulrich, 1984; Warring, 1984a,b.)

EXAMPLE 2.1

Specify an appropriate type of pump driven by an electric motor and calculate the power required to pump 8 gpm of water through a pressure rise of 700 psi at ambient temperature. Although specific costs are not to be addressed, choose a lower cost option if more than one type is available. Assume the electric motor and other driving gear have an overall efficiency (η_m) of 90%.

Table 2.1

Data for the Preliminary Selection of Pumps[a]

	< Centrifugal >			< Rotary >			< Reciprocating >		\<Jet\>
	Axial Flow	Radial	Turbine	External Gear	Screw5	Sliding Vane	Piston	Diaphragm	Jet
Max P (bars)	350	350	50	350	350	350	1000	350	350
Min T (°C)	-240	-240	-30	-30	-30	-30	-30	-30	-240
Max T (°C)	500	500	250	400	370	270	370	270	500
ΔP/stage(bar)	2	20	35	200	20	150	1500	70	1
ΔP total (bar)		200							
Max Q (m^3/s)	5	10	1	0.1	0.1	0.1	0.03	0.006	1
Min μ (Pa·s)		0	0	0.001	0.001	0.001	0.001	0.001	
Max μ (Pa·s)		0.2	0.1	400	1000	100	400	100	
Efficiency (%)	50-85[b]	50-85[b]	20-40[b]	40-85	40-70	40-85	60-90	40-70	5-20
Relative costs									
Purchase	Low	Low	Mod	Mod	High	Mod	High	Mod	Low
Installation	Low	Low	Mod	Low	Low	Low	High	Mod	Low
Maintenance	Low	Low	Mod	Low	Mod	Low	High	Mod	Low
Service compatibility[c]									
Cavitation	D	E	B	B	B	B	B	B	A
Corrosive	C	C	C	C	C	C	C	C	A
Dry operation	E	E	E	D	D	E	E	B	A
High flows	A	A	E	D	D	D	D	E	E
High pressures	X	C	B	B	B	B	A	B	X
High temperatures	C	C	D	C	C	C	C	C	A
Low flows	X	D	A	A	A	A	B	A	B
Variable flows	A	A	B	C	C	C	C	C	A
Variable ΔP	E	D	D	C	C	C	C	C	D
Potential problems[c]									
Pulsations	A	A	A	B	A	B	C,D	C	A
Noise	A	A	A	A	B	B	D	B	B
Reversibility of flow	X	X	X	C	X	X	C	C	X
Overpressure protection	A	A	D	C	C	C	C	C	A
Other									
Disadvantages			d,e	d,e	d,e	d,e	d,e	d-f	g
Advantages							h	h	

Footnotes: **a.** Modified from Ulrich (1984). Used with permission. **b.** Independent of viscosity up to 0.05 Pa·s. **c.** Key: A=excellent, B=modest limitations, C=special units available at higher cost to minimize problems, D=limited in this regard, E=severely limited in this regard, X=unacceptable. **d.** Motor gear reducers are often necessary. **e.** Pressure relief protection necessary. **f.** Diaphragm failure should be anticipated. **g.** Process fluid may be contaminated by motive fluid. **h.** Operated conveniently with steam or compressed air.

To solve this, first consider the various types of pumps listed in Table 2.1. Since nothing specific is noted in the example statement here about service, assume no special consideration is required. Next, note that a ΔP of 700 psi is given. Converting this to SI, the pressure rise at 48.23 bar (4823 kPa) is too large for some of the pumps shown. While it may be possible to install more than one pump in series and achieve the necessary pressure differential, often this complicates the installation and maintenance situation. Hence, eliminate all pump categories that do not have a high enough single-stage pressure increase for the given situation. This leaves the following candidates: external gear, sliding vane, reciprocating piston, and reciprocating diaphragm.

Next, consider the cost aspect. The reciprocating piston type is preferred at very high ΔP's, but at this moderate value this type pump may have a premium first cost. This is demonstrated by the "high" notation in the purchase cost category. All of the other three have "moderate" purchase costs. Since the diaphragm pump has a moderate maintenance cost and the other two are "low," eliminate the diaphragm model. In checking the remaining two--the external gear and the sliding vane--it is difficult to eliminate one relative to another. For this reason, either will suffice for this exercise. Arbitrarily choosing an efficiency value from the range given there (both types have 40%-85% given), assume that $\eta_p \approx 70\%$. Now the power required can be calculated.

From basic thermodynamics considerations, the power can be calculated for a steady-state, steady-flow adiabatic process from the mass flow rate and an integral involving the pressure and the specific volume. Assuming further that the specific volume does not change significantly with pressure gives (V is the volume flow rate)

$$W = m \left(-\int v\, dP\right) \approx - m\, v\, \Delta P = - m\, (\Delta P)/\rho = - V\, \Delta P$$

Incorporating both the pump and motor efficiencies will allow the electrical power required to be found:

$$W \approx -V\, \Delta P / (\eta_p\, \eta_m)$$

Substituting the appropriate numbers yields the following result:

$$W \approx \frac{-(8 \text{ gal/min})(0.003785 \text{ m}^3/\text{gal})(700 \text{ psi})(6.894 \text{ kPa/psi})}{(60 \text{ min/s})(0.7)(0.9)} = -3.87 \text{ kW}$$

23

2.3 FANS

Fans share some general characteristics with pumps. Three categories of fans are normally defined. These are *axial, propeller,* and *centrifugal.* Applications are found in numerous locations in an industrial plant. Included are forced- or induced-draft cooling towers, blowers, ventilators, air-conditioning fans, and other places where there is need to move a gas, particularly air. Factors normally considered in applications are volume handled versus pressure rise across the fan and the ability to control the flow rate.

Axial and propeller fans are closely related in that both move air by the angle of attack of the rotating blade. In the axial fan, the housing around the blade has an important interaction with the gas, particularly in control functions. In the propeller-type fan, the housing plays little or no role in controlling the flow. Hence, propeller fans are not desirable for control purposes. As in the centrifugal pump, the centrifugal fan imparts a significant amount of kinetic energy to the gas during the latter's outward flow through the fan. The centrifugal force acting at the perimeter of the fan causes a pressure rise as the gas exits.

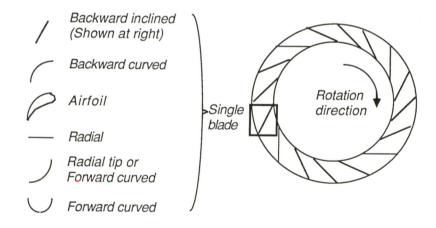

Figure 2.6 Examples of some blade profiles found on fans.

Specific fan characteristics are, of course, important in determining final designs. For preliminary designs, they are less critical. To touch on just a few aspects of fan specification, consider the examples shown in Figure 2.6. Typically, the comments below apply to centrifugal fans, but similar considerations are found in other types also. *Backward-inclined* fans are generally less expensive than some of the other types available. They do, however, demonstrate an instability over a range of flow rates. This type of fan has the characteristic that it can never be overloaded because of increases in load. The *backward-curved* fan shows similar characteristics, but the instability region is less pronounced. Both of these fans are used at high speed so that service in streams with suspended particles is not recommended due to possible erosion.

Table 2.2

Data for the Preliminary Selection of Fans[a]

	<	Centrifugal	>	<	Axial	>
	Radial (paddle wheel)		Backward Curved (squirrel cage)	Tube	Vane	
Absolute P (gauge, bars)	Near 1		Near 1	Near 1	Near 1	
ΔP/stage	15 kPa		10 kPa	1 kPa	5 kPa	
Max Q (std m^3/s)	300		500	300	300	
Efficiency range (%)	65-70		75-80	60-65	60-70	
Relative costs						
Purchase	Moderate		Low	Low	Moderate	
Installation	Moderate		Moderate	Low	Low	
Maintenance	Low		Low	Low	Low	
Compatibility[b]						
Corrosive gases	C		C	C	C	
High-temperature gases	C		C	D	D	
Abrasive gases (particles)	A		C	D	D	
Vacuum service	X		X	X	X	
Variable flows	A		A	A	A	
Variable pressures	E		E	E	E	
Performance problems[b]						
Lubrication contamination	A		A	A	A	
Flow pulsations	A		A	A	A	
Noise	B		B	D	D	
Vibration	B		B	A	A	
Explosion hazards	B		B	B	B	
Other					Flow direction reverses easily	

Footnotes: **a.** Modified from Ulrich (1984). Used with permission. **b.** Key: A=excellent, B=modest limitations, C=special units available at higher cost to minimize problems, D=limited in this regard, E=severely limited in this regard, X=unacceptable.

Use of the airfoil type of blades is a way of increasing fan efficiency. This comes with an increased cost compared to that of flat-blade devices. Usually, airfoil blades are applied to clean gas streams to minimize blade erosion.

Radial blades usually form one of the least cost fan types. Blade speed is slow, and the efficiency is low. High particulate loadings in the gas stream is a situation where this fan is often used. Blades are most easily replaced on open configurations (no shroud around the blades), and these types are often found in highly erosive environments. Radial-tip blades may be a more desirable configuration for erosive environments as

25

operating costs are lower than the straight-blade form. The forward-curved blades produce large volume flows for a given fan size. They are not recommended for potentially erosive flows.

Fan performance characteristics are a function of the type of fan used. Many types of fans yield characteristics that are not too different than those shown for pumps in Figure 2.4. In applications, it is desirable to consider control factors as well as whether or not the fan is self-limiting in case of an unanticipated addition of load resistance.

Some criteria that should be considered when choosing fans are given in Table 2.2. An excellent summary of most aspects that should be considered in the application of many of the fan types available is found in the literature (B&W, 1972; Singer, 1981; Reason, 1983; Thompson and Trickler, 1984).

2.4 COMPRESSORS

Compressors overlap with both pumps and fans in terms of physical characteristics. Typically, compressors are taken to be devices that perform a pressure increase process on gases where the total pressure change is larger than that found from fans. The latter devices are normally meant to cause the gas to flow without sizeable pressure increases. Partially like some aspects of pumps and some aspects of fans, compressors are found in reciprocating, centrifugal, and axial flow types. Since the general characteristics of these types of operation have been discussed above, they will not be repeated. Much of what does appear here is adapted from Dimoplon (1979).

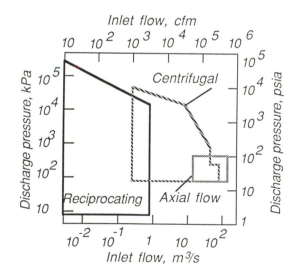

Figure 2.7 Approximate ranges of reciprocating, centrifugal, and axial flow compressors. Adapted from Dimoplon (1979).

The ranges of applicability of the various types are important. Dimoplon (1979) has given a set of ranges for these types of compressors. See Figure 2.7.

As is the case with pumps, the *reciprocating* types of compressors are favored for lower flow, higher pressure service compared to other types of compressors. Usually, higher pressure service is accomplished in stages, with maximum pressure ratios of 3 or 4 found in each stage, and with a maximum pressure rise in each stage limited to about 1000 psi (6900 kPa). Overall adiabatic efficiencies for these devices can range from 60% to 80%, but they are typically in the range of 75% to 80%.

Centrifugal compressors are found in a variety of applications. The reasons for this involve both their operational characteristics and ease of repair. Percent polytropic efficiencies of centrifugal compressors are usually in the 70% to 80% range.

Axial flow compressors are found in a more limited range of output conditions, both in terms of flow rate and pressure rise, than are other types of compressors. Axial flow types overlap somewhat with centrifugal compressors in their overall operational ranges. These devices are favored over the centrifugal types when higher operating efficiencies are desired. Efficiencies of axial flow compressors are in the range of 80% to 85%.

Selection data is given in Table 2.3, and additional information on compressors is available in the literature. (See, e.g., Gulf, 1979; Bloch et al., 1982; Brown, 1986.)

2.5 TURBINES AND OTHER EXPANDERS

There are many situations where turbines and other expanders may be required in a design application. Some of these cases obviously include power plant design. In addition, however, there may be occasions where a pressure drop is taken in some other type of device. You as the engineer have the option of designing in a totally irreversible pressure drop (i.e., across a valve) or going for a partially reversible pressure drop through some type of work-producing expander.

Turbine selection for steam power plants is a highly refined business. In practice, most of this work is performed in conjunction with a power plant equipment vendor, primarily General Electric and Westinghouse in this country. Company representatives have presented papers on various aspects of design considerations over the last several decades. Reprints of this information are available from the vendors. See, for example, Bailey et al. (1967), which has been reprinted by General Electric.

Not so prevalent, compared with information on steam turbines, is application on turbines for other fluids and other types of expanders. Limited information on general expanders has appeared in the literature over the last two decades. One example is the text by Bloch et al. (1982). Use Table 2.4 to assist you in making selections between types. Note that the term *turndown ratio* used in that table refers to the fraction of full flow rate that can still operate the turbine or expander.

Table 2.3

Data for the Preliminary Selection of Compressors[a]

	< Centrifugal >		< Axial >		< Rotary >			<Recip>	<Other>
					Twin-Lobe Single & Staged	Screw Single & Staged	Slidng Vane Single & Staged	Piston Single & Staged	Ejector Single & Staged
	Single	Staged	Single	Staged					
P range (atm)	.1-2	.1-700	.1-2	.1-14	.3-2	1-10	.1-10	.01-3E3	.01-5
Max Stage P_2/P_1	1.4	1.2	b	1.4	2.0	4.0	4.0	4.0	-
Max Stages		8		15	1	1	1	8	5
Max Q (std m^3/s)	80	200		300	20	15	0.8	1.5	-
Efficiency, %	70-80 50-70[c]	70-80 50-70[c]	80-85 50-70[c]	80-85 50-70[c]	60-80 40-60[c]	60-80 40-60[c]	60-80 40-60[c]	60-80	25-30
Relative costs									
Purchase	Mod	Mod	Mod	High	Mod	Mod	Mod	High	Very low
Installation	Mod	Mod	Mod	High	Mod	Mod	Mod	High	Low
Maintenance	Low	Low	Mod	Mod	Mod	Mod	Mod	High	Very low
Compatibility[d]									
Corrosive gases	C	C	E	E	E	D	E	D	A
High-T gases	D	D	D	D	D	D	D	E	A
Abrasive gases	C	C	E	E	E	D	E	X	A
Vacuum service	C	C	C	C	B	B	B	A	A
Variable Q's	C	C	E	E	C	C	C	D	A
Variable P's	D	D	E	E	A	A	A	A	A
Problems[d]									
Lube contamination	A	A	A	A	C	C	C	C	D
Flow pulsations	A	A	A	A	C	B	B	C	C
Noise	D	D	D	D	B	C	B	D	B
Vibration	D	D	D	D	A	A	A	B	A
Explosion hazards	D	D	D	D	B	B	B	E	A

Footnotes: **a.** Modified from Ulrich (1984). Used with permission. **b.** Seldom used without staging. **c.** Vacuum operation. **d.** Key: A=excellent, B=modest limitations, C=special units available at higher cost to minimize problems, D=limited in this regard, E=severely limited in this regard, X=unacceptable.

Approximate efficiencies for all types of process expanders (e.g., steam turbines and gas and liquid expanders) is given as a function of output power (P_{out} in kW) in Equation 2.1. This equation is an approximate curve fit of a recommended plot given by Ulrich (1984).

$$\eta_e \approx 0.45 \left(P_{out} / 100 \right)^{0.125} \qquad 100 \text{ kW} \leq P_{out} \leq 10,000 \text{ kW} \qquad (2.1)$$

Use this relationship for preliminary prediction of the efficiency of a given turbine or other expander.

Table 2.4

Data for the Preliminary Selection of Turbines and Expanders[a]

	< Drives >		< Power Recovery Machines >		
	Steam Turbines (noncondensing)	Air Expanders	Gas Expansion Turbines Axial	Radial	Liquid Radial Expanders
Maximum capacity, P_{out} (kW)	15,000	-	5000	1000	1000
Normal feed temperature (°C)	400	25	<500	<550	25
Normal feed pressure (bar)	45	4	<175	<175	b
Normal exhaust temperature (°C)	150		100	Various	25
Normal exhaust pressure (bar)	4.5	1.1	Various	Various	1.1
Turndown ratio	0.6	0.5	0.8	0.7	0.6
Max liquid in discharge (%)	<20		<20	<20	
Efficiency (%)	c		c	75-88	50-60
Compatibility[d] Corrosive fluids Explosive atmospheres	D B	A A	D B	B B	C A

Footnotes: **a.** Abridged from Ulrich (1984). Used with permission. **b.** Liquid expanders can tolerate any pressure that can be contained by a centrifugal pump. **c.** See Equation 2.1. **d.** Key: A=excellent, B=modest limitations, C=higher cost units available to minimize this problem, D=limited in this regard.

2.6 STORAGE VESSELS

Storage vessels are often incorporated into thermal systems to buffer the flow mass or energy of one stream relative to the flow of another. For example, coal may be delivered to a plant once a day, but the coal might be needed on a regular basis throughout the day. The normal solution is to incorporate storage between the train and the plant. This particular storage may simply be a *pile* of coal in the yard, sometimes denoted by the term *open yard storage*. We consider a solid here to be a special case of fluids.

Another common type of storage is the household water heater. Whether this is gas or electrically heated, there is still a need to allow the domestic hot water sufficient time to become fully heated as it travels from the water supply line to the end use. This is needed because there may be times when the utilization of energy via the hot water occurs at a faster rate than the addition of energy via the burner or electrical heater element, or vice versa. Keep in mind that storage can be used for mass or energy accumulation purposes.

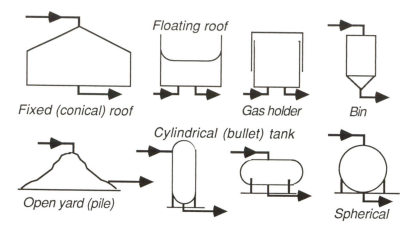

Figure 2.8 Schematic diagrams of common types of storage vessels and forms.

Figure 2.8 shows a variety of types of storage vessels as well as the open yard (pile) type of configuration. Table 2.5 presents a brief summary of factors that may be of value to consider when designing with storage. Of course, this figure and table do not cover all possible types of storage methods and containers, but they do give a fair idea about some of the more common applications.

2.7 VALVES

Valve selection is not normally thought to be of much concern in preliminary design studies, and usually this is indeed the case. On the other hand, valves are often very important in the final design of a system. Although each valve has a small cost compared to most system costs, there may be thousands of valves in a given design; and thus their costs are important. The appropriate choice of a valve could have far-reaching implications about the control of a system. Another point to note is that the choice of a valve type could have important implications about the type of load applied in a pump or compressor circuit.

As has been pointed out by Miller (1983), the use of valves dates to 2700 BC, when designs of leather and wood were used in a "flap" configuration on the island of Crete. A form of a check valve was used by the Romans fifteen centuries before the birth of Christ. Bronze valves of the plug cock variety were used by the Greeks in the construction of fountains at least as far back as 400 BC. Valves have played a very important role in more recent industrialization, including the development of the steam engine. It is imperative that any system designer has at least a basic understanding of the general kinds of valves available and their corresponding application trade-offs.

Table 2.5

Data for the Preliminary Selection of Storage Vessels[a]

	< Atmospheric Storage >					< Pressurized Storage >	
	Fixed (conical) Roof	Floating Roof	Gas Holder	Bin	Open Yard (pile)	Cylindrical (bullet) Tank	Spherical Tank
Typical Maximum Size							
Volume (m^3)	100,000	100,000	20,000	4,000	200,000	1600	15,000
Height or length (m)	15	15	30	50	50	20	30
Diameter or width (m)	90	90	30	10	120	10	30
Length/diameter	<2	<2	1-2	2-5	0.4	2-5	1
Stored Medium							
Solid				X	X		
Liquid	X	X				X	X
Gas			X			X	X
Orientation							
Axis vertical	X	X	X	X	X	X	-
Axis horizontal						X	-
Maximum pressure, (bars, gage)	0.2	0.2	0.2	-	0	17	14
Temperature range (°C)	-20 to 40	-20 to 40	-20 to 40	-20 to 40	-20 to 40	b	b
Common Construction Materials							
Carbon steel	X	X	X	X		X	X
Concrete	X	X		X	X		
Plastics or fiberglass				X		X	
Alloys and coated or clad steel	X	X	X			X	X

Footnotes: **a.** Modified from Ulrich (1984). Used with permission. **b.** For steel, the range is -20 to 600°C; for aluminum, -250 to 200°C; for stainless steel, -250 to 800°C; for nickel-based alloys, -200 to 700°C.

Valves and their performance have been described in numerous articles previously. (See, e.g., O'Keefe, 1971; O'Keefe, 1976a,b; Miller, 1983.) Only some brief highlights regarding valve selection will be given here.

Globe valves (see Figure 2.9) are often used for tight shut-off and throttling situations. The active portion of the valve can be a circular plug that enters a circular hole (the *seat*). In the most-used design, the fluid must turn 90° to flow through the seat and then turn back again to the original direction. As a result, the pressure drop across the valve is relatively high. *Angle valves* have similar characteristics but are used in place of a valve plus an elbow in a piping run. In this manner, the overall cost of the angle valve installation may be less than the cost of a straight-run valve plus the elbow. Both types of valves demonstrate good throttling performance and seal tightly in the "off" position. If wear of the seat does occur, repair is usually easily accomplished compared to many other types of valves.

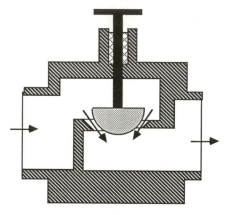

Figure 2.9 Schematic diagram of a globe valve.

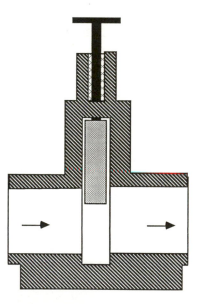

Figure 2.10 Schematic diagram of a gate valve.

Gate valves (see Figure 2.10) operate on a "slide-past" principle, often using a tapered wedge to close off the flow. Open operation usually demonstrates lower pressure drop than a corresponding globe valve. However, the gate valve is not designed for throttling service, as "wire-drawing" erosion can occur to the seat. Frequent "off"-and-"on" service can cause similar problems. The valve is not "bubble-tight" in the off position, and this characteristic is made worse with handling of dirty fluids. Costs of gate valves are usually less than corresponding globe valves.

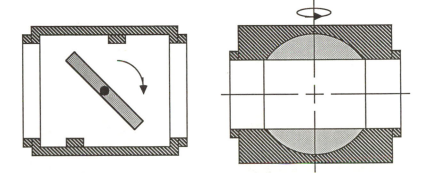

Figure 2.11 Schematic of the active portions of a butterfly valve (left) and a ball valve (right). Arrows here show direction of closure.

For quick off-on operation, as well as low pressure drop in the on position, the *butterfly* and *ball* valves are prime candidates. These devices typically require only a 90° turn of the stem for complete off-on operation. (In fact, the on-to-off action can take place so quickly, there should be some concern shown for waterhammer possibilities.) A schematic of two basic types of off-on valves is shown in Figure 2.11. In the first of these, a 90° rotation of flap arrangement either seals against the shoulders or opens fully to flow. The second of the schematics shows the ball-type construction. This latter design seals better in the closed position and demonstrates lower pressure drop in the open position than the butterfly type, but there is usually a premium cost for the ball type. The *plug valve* (not shown) is schematically similar to the ball valve, but the ball is replaced with a more cylindrically shaped rotating element. Most of the valves in this category have the characteristic that they may not be suitable for high-temperature applications because of the seating material's temperature limitations.

Check valves are used to ensure that the flow in a pipe is in one direction only. This kind of valve is one of the few types that is truly automatic in its operation. The modes of operation of these valves vary from a flapper arrangement like that shown in the left-hand side of Figure 2.12 and a diaphragm concept like that shown in the right-hand side of the figure to a ball stopper arrangement that is not shown here. Most of the valves in this category are not normally considered to be able to react to high-frequency variations in flow, such as might be found in the output of a reciprocating pump. Another concern normally shown about these types of valves is that they are not typically bubble-tight.

A large category of valves found in typical designs is denoted as *control valves.* Whereas many of the basic types of control valves are not too different than some of those discussed above, the emphasis here is on the ability to control the flow rate to desired values. Several possibilities are available in a given design. Some valves open quickly (relative to their total stem travel) while others open more slowly. This has implications on the corresponding pressure change in the line downstream, which is obviously related to the flow rate.

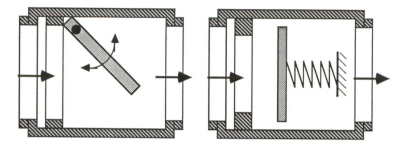

Figure 2.12 The method of operation of two types of check valves is shown. The left-hand side uses a flapper arrangement to constrain the flow to one direction, while the right-hand side uses a diaphragm concept. The normal flow direction is as shown by the horizontal arrows.

See Figure 2.13 for some typical examples of control valve action. Although the overall control of flow depends upon both the effect of the valve and the complete downstream resistance, the valve denoted in the figure as "equal percentage" will often give an approximately linear system characteristic. Most control valves demonstrate performance between this type of variation and the one denoted as "linear" in the figure. Special concern is given in the design of control valves (particularly the *trim,* which is a name for the orifice formed between the gate or plug and the seat) and the material specifications to result in a device with high reliability.

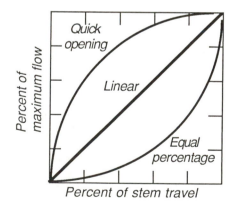

Figure 2.13 Plot of typical responses of various types of control valves

Numerous aspects related to the specification of the many types of valves have not been covered here. In applications, concern would have to be given to the valve (and its various subcomponents') materials, the bonnet design, the body-bonnet connections, the body end connections (this is related to the discussion of piping given in the next subsection), the seating details, the packing, and, of course, the cost. To show the wide variation in the latter, prices for one size of different types of valves are shown in Table 2.6. It should be obvious that valves can vary widely in price and can affect total plant cost. For an introduction to many of the other aspects regarding valve specification, refer to the references at the end of this chapter.

Table 2.6

Prices of Various 4-in. Valves[a]

Description	Price[b]
Class 125 cast iron gate valve bronze mounted	$125
Class 125 cast iron check valve bronze mounted	110
Class 125 cast iron globe valve bronze mounted	250
Class 300 cast steel gate valve 13% chrome steel trim	410
Class 300 cast steel check valve 13% chrome steel trim	450
Class 300 cast steel globe valve 13% chrome steel trim	620
Class 150 ductile iron plug valve with nonlubricated plug	370
Class 300 ductile iron plug valve with nonlubricated plug	550
Class 300 cast steel plug valve with lubricated plug	1050
Class 150 cast steel ball valve with Teflon trim	615
Class 300 cast steel ball valve with Teflon trim	715
Class 600 pipe line valve with cast steel high-temperature trim	7400
High performance butterfly valve with cast steel seats	500

a. From Miller (1983). b. Mid-1983 prices. Quantity purchased and country of origin will affect prices.

2.8 PIPING

Precise specification of piping is almost never required in the preliminary design analysis of systems. But there may be times when there is a need to define some general aspects about piping. Three items should be of special concern when this happens.

One of these has to do with the bore of the fluid channel. This is normally found in a specification of the *schedule.* Schedules of 40, 80, 160, or XX pipe are common.

A second item to be specified relates to the mechanical strength of the fittings. Here the normally found values are 150-, 3000-, and 6000-lb ratings. Usually, the 150-lb components are either cast or forged, while the higher rating counterparts are typically forged. Flanges (see additional discussion below) are found in pressure classes of 150, 300, 400, 600, 900, 1500, and 2500 lb.

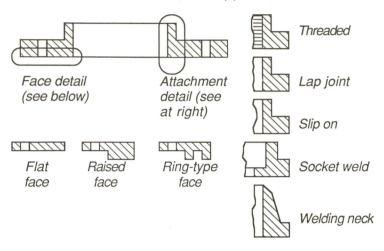

Face detail
(see below)

Attachment
detail (see
at right)

Flat
face

Raised
face

Ring-type
face

Threaded

Lap joint

Slip on

Socket weld

Welding neck

Figure 2.14 Schematic details of the various face designs and attachment forms of common flanges. See accompanying text for additional explanation.

The third item to address is the type of fitting connections to be used. Of course, there are fittings like ells, tees, couplings, and so on, that must be specified; and the correct incorporation of these components into a piping configuration is assumed. On the other hand, some choices can be made regarding the connections. In terms of connections, the normally available options are: threaded, socket welded, and butt welded. Flanges can be used or not used with any of these types of connections, and the flanges are among the most expensive of any elements other than valves. Both the facing of the flange and the general attachment mode are necessary in the specifications. See Figure 2.14.

Some description of the standard attaching designs may be of value. The *threaded* flange has a diameter that matches the pipe size. It is used for low-pressure systems and where welding could be hazardous. The threaded type is not normally used with larger diameter pipe. A *lap joint* type is bored slightly larger than the OD of the pipe, and the radius on the bottom matches the radius on the stub end. The fitting is slipped over the pipe, and the stub end is welded onto the pipe and bolted up. This design is used for systems that need frequent cleaning or inspection.

The *slip-on* flange has a low hub and is bored slightly larger than the OD of the pipe. This flange is welded on both the inside and outside of the flange face to prevent leakage. A welding-neck flange (see discussion that follows) is used more frequently than this design in common application, but the slip-on flange finds usage when cost or space may be a major consideration.

Socket-weld flanges are bored to the ID of the pipe and counter-bored slightly larger than the OD of the pipe to allow the pipe to be inserted and welded in place. Normal application finds this type of flange in smaller diameter and higher pressure systems.

By far the most common flange type is the *welding neck*. It is bored to the ID of the pipe and has a high neck. Use of this type of flange usually results in the soundest welds.

Not shown is the *blind* flange. This is simply a solid circular piece of metal that is drilled for flange bolts. It is used to blank off a piping run.

Special problems are often faced in specifying proper expansion capabilities for piping runs. Two approaches that are considered are piping loops and special expansion joints. As outlined by Broyles (1985), there are numerous reasons why expansion joints are specified, including: limitations on space available or piping diameter too large to make loops practical; desire to minimize pressure drop; better ability to handle abrasive fluids; and pipe diameter. Piping loops may also be favored for a variety of reasons. Included here are: the ability to incorporate loops with needed changes in direction of the piping run; impracticality of expansion joints for very high pressure and/or corrosive service; and possible code violations for the use of expansion joints.

Excellent descriptions of proper piping practices are available. (Included are King 1979; Crane, 1979.)

REFERENCES

B&W, 1972, STEAM/ITS GENERATION AND USE, The Babcock & Wilcox Company, New York.

Bailey, F., K. Cotton, and R. Spencer, 1967, "Predicting the Performance of Large Steam Turbine-Generators Operating with Saturated and Low Superheat Steam Conditions," American Power Conference Paper (reprinted by the General Electric Company).

Bloch, H., J. Cameron, F. Danowski, R. James, J. Swearingen, and M. Weightman, 1982, COMPRESSORS AND EXPANDERS, Marcel Dekker, New York.

Brown, R., 1986, COMPRESSORS--SELECTION & SIZING, Gulf Publishing Company, Houston.

Broyles, R., 1985, "Pipe Loop or Expansion Joint," Chemical Engineering, October 14, pp. 103-106.

Crane, 1979, FLOW OF FLUID THROUGH VALVES, FITTINGS, AND PIPE, The Crane Co., New York.

Dimoplon, W., 1979, "What Process Engineers Need to Know About Compressors," in COMPRESSOR HANDBOOK FOR THE HYDROCARBON PROCESSING INDUSTRIES, Gulf Publishing Co., Houston, pp. 1-8.

Gulf Publishing Co., 1979, COMPRESSOR HANDBOOK FOR THE HYDROCARBON PROCESSING INDUSTRIES, Gulf Publishing Co., Houston.

Hicks, T., and T. Edwards, 1971, PUMP APPLICATION ENGINEERING, McGraw-Hill, New York.

Karassik, I., W. Krutzsch, W. Fraser, and J. Messina (Eds.), 1976, PUMP HANDBOOK, Mc Graw-Hill, New York.

King, R., 1979 (Ed.), PIPING HANDBOOK, FIFTH ED., McGraw-Hill, New York.

Miller, R., 1983, "Valves: Selection, Specification, and Application," Heating/Piping/Air Conditioning, October, pp. 99-118.

Neerken, R., 1974, "Pump Selection for the Chemical Process Industries," Chemical Engineering, February 18, p. 104.

O'Keefe, W., 1971, "Valves," Power, March, pp. S·1-S·16.

O'Keefe, W., 1972, "Pumps," Power, June, pp. S·1-S·24.

O'Keefe, W., 1976a, "Control Valves, Actuators, Regulators, Positioners," Power, April, pp. S·1-S·16.

O'Keefe, W., 1976b, "Check Valves," Power, August, pp. 25-36.

Peters, M., and K. Timmerhaus, 1980, PLANT DESIGN AND ECONOMICS FOR CHEMICAL ENGINEERS, 3RD EDITION, McGraw-Hill, New York.

Pollak, F. (Ed.), 1980, PUMP USERS' HANDBOOK, Gulf Publishing Co., Houston.

Reason, J., 1983, "Special Report--Fans," Power, September, pp. S·1-S·24.

Singer, J. (Ed.), 1981, COMBUSTION POWER SYSTEMS, THIRD EDITION, Combustion Engineering, Inc., Windsor, CT.

Stewart, H., and T. Philbin, 1984, PUMPS, The Bobbs-Merrill Co., New York.

Thompson, J. E., and C. J. Trickler, 1983, "Fans and Fan Systems," Chemical Engineering, pp. 46-63, March 24.

Ulrich, G. E., 1984, A GUIDE TO CHEMICAL ENGINEERING PROCESS DESIGN AND ECONOMICS, Wiley.

Walker, R., 1972, PUMP SELECTION, Ann Arbor Science Publishers, Ann Arbor, Michigan.

Warring, R., 1984a, PUMPING MANUAL, SEVENTH ED., Gulf Publishing Co., Houston.

Warring, R., 1984b, PUMPS: SELECTION, SYSTEMS AND APPLICATIONS, SECOND ED., Gulf Publishing Co., Houston.

PROBLEMS

2.1 Size and specify a pump for the following application. A process fluid is to be pumped at a rate of 100 gpm through a pressure rise of 100 psi. Your company lab measures the density and the viscosity and reports their values as 52 lbm/ft^3 and 2000 centipoise, respectively. Cost, although not a totally dominating factor, should be made no larger than necessary. Be sure to state any considerations you make when selecting an appropriate pump. This should include not only why particular pumps are rejected and one selected, but it should also include any potential application problems with the one selected (e.g., if flow reversal could be a problem with the pump selected, so state).

2.2 A fan is to be selected for installation in your manufacturing facility. In this facility, construction of heavy equipment is performed, including cutting, welding, sand blasting, and use of large cranes. There are many particulates in the air around the fan inlet. It is desired to move 60,000 cfm from the facility (atmospheric pressure is 12.9 psia, temperature is usually around 80°F) through a pressure rise of 0.2 psi. Specify and size a fan for this application. Be sure to state any considerations you have through the selection process.

2.3 The preliminary design of a small power plant is being performed. One required aspect of this design is the specification of the fan for the boiler. Design output of the plant is to be 100 kWe. The overall plant efficiency is anticipated to be approximately 38%. Natural gas will be the fuel used. It is assumed that a boiler can be configured such that it will have an efficiency of 91% and a pressure drop at design conditions of approximately 4 kPa. Select and size a fan for this application. Make appropriate assumptions if necessary.

2.4 A volume flow rate of 17 m^3/s of air is required at 14 bars, and this air is to be furnished by a new compressor from atmospheric conditions in a plant (approximately 0.95 bars and 30˚C). Determine a power requirement for this compressor assuming an appropriate efficiency, find the output temperature of the air, and specify an appropriate type of compressor for this duty.

2.5 The feedstream of a fluid (specific heat of 2.1 kJ/kg˚C, density of 1100 kg/m^3) for a process plant must be heated from 20 to 65˚C. It is anticipated that the fluid will flow at a rate of 20 kg/s for periods of 3 min and then no flow will occur for 10 min. Size a heater tank (much like a home water heater) that uses electricity to accomplish the heating. Specify the physical size of the tank and the electrical element rating required such that the size of each is minimized. Add an "over-design factor" to the specification (safety factor) by assuming that the flow rate of the fluid is actually 30 kg/s.

2.6 Contact a pump vendor and secure information of the sort shown in Figure 2.4 for a single product line of pumps. Plot the information if that is not already done.

2.7 From information similar to that used in Problem 2.6, but for only a single diameter, present appropriate curves of pressure rise versus flow capacity for two of the pumps connected in series and in parallel.

2.8 Summarize the key trade-offs between radial and axial flow gas expansion turbines.

2.9 Air is to be compressed from atmospheric conditions to an outlet pressure of 3 bars (absolute). Make up a brief table that shows the inlet flow ranges of the appropriate kinds of compressors that could be used.

2.10 CO_2 is available at one point in a process at 40 bars and 300˚C. It is desired to estimate the power that may be available if this stream is expanded through an appropriate work device to atmospheric pressure. A flow rate of 100 kg/s of this gas is available. Select an appropriate expander, estimate the power output, and determine the temperature of the exhaust.

2.11 Analyze a jet pump to move water at 200 kPa to a pressure of 300 kPa. A total of 40 kg/s of water at 40°C is needed. Plot the pressure against the flow rate (both for the high-pressure stream) for this situation. Make and state appropriate assumptions about the conversion of the high-pressure stream in the pump.

2.12 Make a list of the important considerations that would go into a choice between an axial flow compressor and a centrifugal compressor.

2.13 Specify an appropriate type of pump driven by an electric motor, and calculate the power required to pump 800 gpm of water through a pressure rise of 80 psi. Choose a lower cost option if more than one type is available. Assume that the electric drive and associated hardware have an overall power conversion efficiency of 89%.

2.14 By referring to some of the literature cited in this chapter or some other appropriate information, make a list of factors that must be specified for the actual purchase of a centrifugal pump.

2.15 List the considerations that must be given for the actual specification of an axial flow compressor. You may consult a vendor, reference material, or whatever. Be sure that your list is reasonably complete.

CHAPTER 3

HEAT EXCHANGE
DESIGN OPTIONS

3.1 INTRODUCTION

Consider now various design options in heat exchange. A major element in this general topic is heat exchangers. These devices can be found in so many configurations that a person who has been simply introduced to the Log-Mean-Temperature Difference and Effectiveness-NTU methods of analysis can be quite perplexed in trying to determine which of the almost limitless types of heat exchangers available, many apparently satisfying the required heat transfer duty, should be used.

Designs that incorporate tubes, for example, are only a subset of the many heat exchangers available. In spite of being only a subset, there is an organization that sets standards for tubular heat exchangers (TEMA). [Actually several organizations deal with standards for heat exchangers including Tubular Exchanger Manufacturers Association (TEMA, 1978), American Society of Mechanical Engineers (ASME, 1980), Heat Exchange Institute (HEI, 1978), and American Petroleum Institute (API, -).] Since heat exchangers can be found in so many forms, distinctions are often left out of discussions in beginning heat transfer texts. However, since heat exchangers find broad applications in virtually all thermal systems, some practical aspects are addressed here. As will be seen, "heat exchangers" is a broad term involving both single devices as well as systems.

In a final section of this chapter, some aspects related to design choices in thermal insulation are discussed. It is not always the case that insulation selection is a critical element in preliminary design studies, but some knowledge of this facet of plant engineering can be of value at that point.

3.2 HEAT EXCHANGERS

3.2.1 Overview

Heat exchangers have been the focus of uncountable articles, papers, and books. Most of these outline the numerous design analysis techniques

41

that are available. (Included are Kern, 1950; Fraas and Ozisik, 1965; Kays and London, 1964; Butterworth and Cousins, 1976; Feldman et al., 1976; Butterworth, 1977; Bellotty and Stock, 1979; Karaç et al., 1981; Karaç et al., 1983; Pettigrew et al., 1983; Shah, 1983a,b; Taborek and Hewitt, 1982; Crane and Gregg, 1983; Boehm and Kreith, 1987.)

Analysis of most heat exchanger applications is done in a manner very much like that discussed in introductory heat transfer texts, either using the conventional modified Log-Mean-Temperature Difference (LMTD) approach or the Effectiveness-NTU (ε-NTU) technique. In general, the LMTD approach may offer a more direct solution if all of the temperatures of a given design are known and other quantities must be found. The ε-NTU technique is preferred when the outgoing temperatures are not known, but this method can be used for any case. In practice, there is a wide variety of methods being applied that range from simply entering the LMTD approach on a programmable calculator (Crane and Gregg, 1983) to detailed hydraulic analysis of the internal flow field in a heat exchanger (Pettigrew et al., 1983).

Extensive tabulations of the functional forms of the ε-NTU method are given by Shah (1983b). He also gives the relationships between the LMTD variables and the ε-NTU variables.

Often, the most critical step in the analysis of a heat exchanger is the determination of the overall heat transfer coefficient, U. This involves the application of convection and/or phase change correlations to find the surface coefficients, h, and use these with the areas, A_1 and A_2, and wall resistance, R_w, to find the result of Equation 3.1.

$$U_1 A_1 = U_2 A_2 = \frac{1}{1/h_1 A_1 + 1/R_w + 1/h_2 A_2} \tag{3.1}$$

The determination of pressure drop should be evaluated also as this quantity is always important in design applications. Some heat exchangers that perform very well thermally may have very high pumping power requirements. Evaluating the relative economic values of heat and work is the focus of a later chapter.

Short of performing detailed studies of various types of heat exchangers to see which type is preferred for a particular application, some shortcut methods have been given in the literature. One such method is the "effectiveness index" technique described by Brown (1986). He defines the effectiveness index as the overall heat transfer coefficient for the given heat exchanger divided by the cost per unit area of the heat exchanger. Of course, the higher the value of this index, all other factors being constant, the better the buy. Average values typically range between 1 and 6 {Btu/hr°F$} for a variety of heat exchangers.

While much design information is given generally in the various monographs on heat exchangers, the beginning design engineer is rightfully perplexed by the myriad of types of heat exchangers available. Which type should be used in a specific application? If only the most appropriate type could be selected, it might not be too difficult to analyze. To outline general characteristics of certain types of heat exchangers, some categorizations of these devices follow. Clearly, the list of types of heat exchangers, as well as each type's advantages and disadvantages, is necessarily limited in scope. An attempt will be made to touch only on some of the more important aspects. A summary of much of the information discussed in the following section is given in Table 3.1. You may wish to refer to this to put the discussion into perspective.

3.2.2 Shell-and-Tube Heat Exchangers

3.2.2.a General Comments

By far the most widely applied heat exchangers are those constructed of a "shell," which contains one of the fluids as well as the tubes, and the "tubes," which contain the other fluid. The heat transfer thus takes place between the two fluids across the tube walls. Classification and discussion of characteristics of these devices have been given in numerous places. (See, e.g., Lord, et al. 1970; Fanaritis and Bevevino, 1976; Gutterman, 1980; Mehra, 1983, as well as the monographs noted above.) For purposes here, it is appropriate to subdivide the shell-and-tube category into three major subcategories: *return bend* (sometimes called *U-tube*), *fixed tubesheet*, and *floating head* (sometimes called *floating tubesheet*.)

Before describing the generally encountered configurations, some information about applications will be given.

◊ Tube Diameter--Make this as small as possible to increase the surface area per unit volume of fluid. Limitations on pressure drop and the ability to clean the outside of tubes may place a lower limit on this parameter.

◊ Tube Length--Generally make this as long as possible to decrease costs. Consider the cost of heat exchange duty per unit length of tubes to see whether or not longer tubes are, indeed, increasing performance.

◊ Tube Pitch--Usually a triangular arrangement will decrease the overall size. Other arrangements (e.g., square) might be needed to decrease pressure drop and allow mechanical cleaning.

◊ Shell Design--Often the fluid on the shell side is in laminar flow. Concern must be given that the shell-side fluid cannot shortcut the desired path, and this is accomplished with well-designed baffles.

To achieve desirable shell-side velocities, multiple shells in series might be used for low flows, and high flows might require the use of multiple shells in parallel.

Table 3.1

Data for the Preliminary Selection of Surface Heat Exchangers[a]

	<	Shell and Tube		><	Plate ><	Air Cooled>
	Fixed Tube Sheet	U-Tube	Bayonet	Floating Head[b]	Flat Plate	Fin-Fan
Maximum surface area (m^2)	800	800	100	1000	1500	2000
Typical number of passes (shell/tube)	1-2/1-4	1-2/2-4	2/1	1-2/1-4	1/1	1/2
Maximum operating temperature (˚C)	150	350	350	350	260	260
Typical maximum operating pressures, bar (shell/tube)	140/140	140/140	140/140	140/140	20/20	-/140
Maximum practical ΔT approach (˚C)	5	5	5	5	1	5
Maximum Flow Capacity Liquid (m^3/s, shell/tube)					0.7/0.7	
Gas (std m^3/s, shell/tube)	15/15	15/15	2/2	15/15		-/15
Typical Mean Flow Velocity (m/s) Liquid (shell/tube)	1-2/2-3	1-2/2-3	1-2/2-3	1-2/2-3	-	-/2-3
Gas (shell/tube)	5-10/ 10-20	5-10/ 10-20	5-10/ 10-20	5-10/ 10-20	-	3-6/ 10-20
Compatibility[c] Fouling service (shell/tube)	E/B	D/D	A/D	B/B	A/A	-/A
Cleanability	E/B	D/D	B/E	B/B	A/A	-/A
In-service tube replacement	A	D	A	A	A	A
Differential thermal expansion	C	A	A	B	A	A
Thermal shock	E	A	A	D	A	A
Toxic or hazardous fluids (shell/tube)	A/A	A/A	A/A	X/A	B/B	-/A
Condensing service (shell/tube)	A/B	B/B	A/B	A/B	E/E	-/B
Evaporative service (shell/tube)	A/A	A/D	A/D	A/A	E/E	-/E
Viscous liquids (shell/tube)	E/B	E/B	D/D	D/B	B/B	-/B
Maintenance	B	D	B	B	A	A
Alloy construction (shell/tube)	D/C	D/C	D/C	D/C	C/C	-/C
Heat transfer efficiency	B	B	D	B	A	B
Relative cost (1 = low, 4 = high)	1	1	4	2	4	2
Pressure drop (bar) Shell	0.2-0.6	0.2-0.6	0.2-0.6	0.2-0.6	0.5-1.5	0.0012
Tube	0.2-0.6	0.2-0.6	0.4-1.0	0.2-0.6	0.5-1.5	0.2-0.6

Footnotes: **a.** Modified from Ulrich (1984). Used with permission. **b.** Packed-tube sheet type. **c.** Key: A=excellent or no limitations, B=modest limitations, C=special units available at higher cost to minimize problems, D=limited in this regard, E=severely limited in this regard, X=unacceptable.

◊ Shell-Side versus Tube-Side Applications of Fluids--Normally, turbulence is more easily initiated on the shell side because of the typically more complicated flow path there. Hence, apply higher viscosity and lower flow rate fluids on the shell side, all other factors being equal. Also, if a fluid use requires periodic cleaning of the heat transfer surface or has other special needs (e.g., if it is fouling, corrosive, toxic, or has a high pressure or a high temperature), it is normally flowed through the tubes. Highest heat transfer per unit of pressure drop is generally possible in the tubes. Condensing fluids are usually placed on the shell side. A need for specialty metals in contact with the condensing fluid may necessitate tube-side condensation to decrease costs.

◊ Counterflow versus Parallel Flow--It is well known that counterflow offers the potential for the maximum temperature change of a fluid stream when compared to parallel flow. Only in special circumstances is counterflow not used, and many of these circumstances are a result of the physical layout of the heat exchanger. Also, if one of the fluids is only changing phase, there is no distinction between counterflow and parallel-flow performance.

For simple shell-and-tube heat exchangers, the tubes represent approximately 60%-75% of the total purchase cost (Dart and Whitbeck, 1980). Hence, the type of material used for the tube construction can have a major effect on the cost of the heat exchanger. Relative costs of some of the more common tube materials are given in Table 3.2.

Table 3.2

Relative Cost of Common Heat Exchanger Tube Materials[a]

Material	Approximate Relative Material Cost
Low-carbon steel	1.0
Copper	1.1
Red brass	1.2
Admiralty brass	1.3
90/10 copper-nickel	1.6
Aluminum	2.0
304 stainless steel	2.5
316 stainless steel	3.0
Nickel	5.0
Monel	5.0
Inconel	8.0
Titanium	12.0
Hastelloy	16.0

[a]From Dart and Whitbeck (1980).

Operating pressures have profound implications on the cost of shell-and-tube heat exchangers also. While the specific application will dictate the details of the costs, an example of pressure effects on shell cost will be noted. This information is given Table 3.3. Similar kinds of cost effects due to tube pressure are also present. Additional cost factors are given in the Appendix.

Table 3.3

Effect of Shell-Side Design Pressure on Heat Exchanger Cost[a]

Design Pressure MPa (psi)	Heat Exchanger Relative Cost
2 (300)	1.0
4 (600)	1.3
5 (750)	1.6
7 (1000)	2.0
8 (1200)	2.5

[a]From Dart and Whitbeck (1980).

3.2.2.b Return Bend (U-Tube) Type Shell-and-Tube Heat Exchangers

One of the more common shell-and-tube heat exchangers in use in industry is the return bend, or U-tube, type. A simplified drawing of one of these devices is shown in Figure 3.1. In practice, many tubes would be used where only one tube is shown in the figure. The vertical plates shown can serve as supports for the tubes, baffles to change the flow direction of the shell-side fluid, or both purposes.

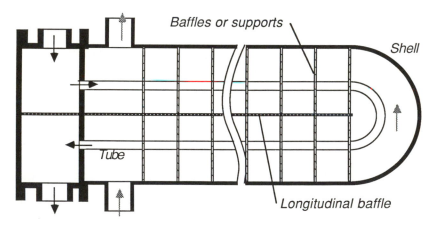

Figure 3.1 A sketch of a simplified shell-and-tube heat exchanger. The shell-side fluid flows as shown by the upward-pointing (gray) arrows, and the tube-side fluid flows as shown by the black arrows. While only one tube is shown here, in practice, a large number of tubes would be used.

The shell-and-tube configuration allows a great deal of flexibility in design applications and, as a result, is frequently found in industrial systems. Among the large number of <u>advantages</u> attributable to this type of heat exchanger are the following: **(i)** large variations in temperature of the service can be tolerated because the tubes can readily expand or contract, and no added expansion joints (as are found in some other types of heat exchangers) are necessary; **(ii)** very high pressures can be applied on the tube side; **(iii)** generally, this type of design is less expensive than many of the other tube-and-shell types; **(iv)** the tube bundle can usually be removed for cleaning (this can be a difficult task, however) or repair (not all tubes may be individually accessible for removal); **(v)** no internal gaskets or special packings are required; and **(vi)** in design configurations that are slightly different than those shown in Figure 3.1, the shell-side inlet can be moved to a point where the shell-side fluid will not impinge on the tubes upon entry. The latter characteristic may be desirable in some applications where the direct impact of the entering fluid could erode away the surface of the tubes at the location of incidence. Alternatively, an impingement plate can be located to protect the tubes from the incoming shell-side fluid, but this is an added expense in the unit.

This type of heat exchanger is not without its <u>disadvantages</u> in some applications. Included are the following: **(i)** typically, service is restricted to clean fluids because cleaning is difficult by mechanical means on either the shell or tube side; and **(ii)** many of the tubes may be almost impossible to replace if a failure occurs. **(iii)** U-tubes always result in even numbers of tube passes. While this is a limitation, it is only minor.

Tube bundles similar to the tube section of a shell-and-tube heat exchanger are often inserted into vessels to serve as stand-by heaters or boilers. This form is sometimes called a *bayonet* application, but this name is also given to a modified form that employs straight tubes, sealed at one end and fixed to a tubesheet at the other. In this latter form, the bundle is inserted into a shell. A second bundle is formed from a second series of smaller tubes, each open at both ends but fastened at only one end in another sheet. The second bundle is inserted into the first bundle with the tubesheets separated by a spool spacer. This latter form of bayonet heater offers some operational and maintenance improvements over the U-tube variety but at an increased cost. The straight-tube form shares some common characteristics with fixed-tubesheet exchangers described in the next section.

3.2.2.c Fixed-Tubesheet Exchangers

A simplified sketch of a fixed-tubesheet heat exchanger is shown in Figure 3.2. In contrast to the U-tube heat exchanger, here the tubes are attached to both sides of the shell arrangement.

The fixed-tubesheet heat exchanger is similar to the U-tube in many respects. One clearly distinguishing feature is that an end chamber is found in this type that functions to return the tube-side flow through the second set of tubes. Hence, the U-portion of the heat exchanger is replaced with a plenum region in the fixed-tubesheet device. This leads to some differences in possible function compared to the U-tube heat exchanger.

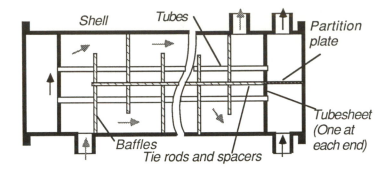

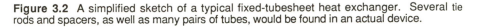

Figure 3.2 A simplified sketch of a typical fixed-tubesheet heat exchanger. Several tie rods and spacers, as well as many pairs of tubes, would be found in an actual device.

Among the <u>advantages</u> of the fixed-tubesheet types of exchangers are the following. **(i)** This type of heat exchanger can handle fouling fluids on the tube side. The straight-through form of the tubes allows mechanical cleaning inside. **(ii)** Fixed-tube exchangers generally allow configurations with odd- (including one) or multiple-tube passes. As noted before, the odd number of passes is not possible with U-tubes. **(iii)** A fixed-tube arrangement has fewer joints than other types of straight-through exchangers. **(iv)** Minimum capital expense is involved in the fixed-tubesheet type when compared to other straight-through types. **(v)** Of all straight-tube types of exchangers, the fixed-tubesheet type offers the maximum protection against leakage of the shell-side fluid to the environment. **(vi)** This configuration can result in the minimum shell diameter of all shell-and-tube heat exchangers for a given heat transfer surface, with the same diameter, length and number of tubes, and tube passes (Mehra, 1983).

Of course, some <u>disadvantages</u> are also inherent in this type of device. **(i)** Perhaps the most important concern is that thermal stresses can become critical if the effects of temperature profiles in the tubes and shell are not matched appropriately. Too much expansion in the tubes compared to that of the shell, for example, will cause the tubes to be in compression. This condition can have very involved implications because all aspects of operation (including start-up, shutdown, normal operation, and any unusual situations) must be considered. Expensive expansion joints may be required. **(ii)** While the tube side can accommodate fouling fluids because mechanical cleaning is easily accomplished there, the shell side is restricted to clean fluids.

3.2.2.d Floating-Head Exchangers

These heat exchangers are a variation of the fixed-tubesheet types, designed to accommodate the movement of the tubes that might result from thermal expansion and contraction. See Figure 3.3. For this reason, these exchangers yield most of the advantages of the fixed type without the drawback of concern for thermal stress effects. In fact, this seemingly simple modification complicates the design and maintenance of the device. As a result, both capital and operational costs are higher. Care must be taken in the design of the way the floating head operates so that leakage does not result. This implies many concerns including the need to make sure that uneven thermal expansion of the tubes does not "tip" the head.

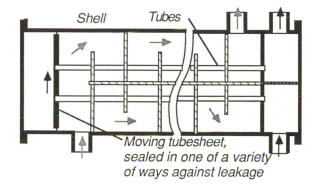

Figure 3.3 A simplified sketch of a floating-head-type heat exchanger. Note that this device is very similar to the one shown in Figure 3.2, except that this one has a movable tubesheet on the left-hand side. The way that the movement of the tubes is accommodated while not allowing the tube-side fluid to mix with the shell-side fluid varies considerably in actual designs.

3.2.3 Plate Heat Exchangers

Plate heat exchangers are a relatively early development, with at least one patent dating back to the late 1870s (Clark, 1974). These devices are made from specially formed *metal plates with grooves* pressed in them similar to the simple schematic shown in Figure 3.4 and the right-hand side of Figure 3.5. The grooves serve two basic functions: aid the heat transfer process and add mechanical rigidity to the overall assembly. As shown in the right-hand side of Figure 3.5, actual plate heat exchangers differ from the one shown in Figure 3.4 in that a given plate could incorporate several hundred grooves and would be built out of several plates stacked together. Heat transfer occurs between two streams across *n* plates in the overall assembly. Two additional plates form the two outer containers of the complete device. Usually, the plates are constructed from a cold-worked metal, often a stainless steel with a thickness around 1 mm. Plate heat exchangers are seldom constructed from mild steel. *Seals,* often referred to as gaskets, are used to contain the fluids within the flow channels of the assembly. The plates are held together in a layered form by a frame and tightening bolts. Plate heat exchangers can be purchased with total heat transfer areas down to a fraction of a square meter and up to over 1000 m^2.

Photographs of a commercial unit and a typical plate are shown in Figure 3.5.

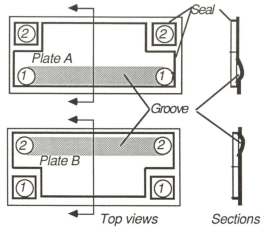

Figure 3.4 Conceptual sketch of two plates from a simplified plate heat exchanger. Plate A would be mounted in front of plate B, and one fluid (1) would flow in and out of one pair of ports, and the other fluid (2) would flow through the second pair.

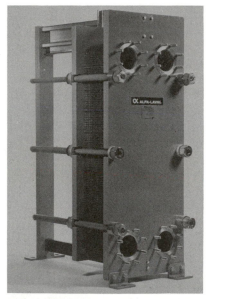

Figure 3.5 A commercial plate heat exchanger is shown on the left, while a typical plate from such a unit is shown on the right. Photos courtesy of Alfa-Laval, Inc.

Performance prediction techniques for plate heat exchangers can be found in a number of papers and books. One such paper has recently compared the prediction techniques available in the open literature (Caciula and Rudy, 1983). In general, these devices have approximately the same effectiveness as a true counterflow heat exchanger.

It is of value to contrast the plate heat exchanger to the shell-and-tube configuration in terms of advantages and disadvantages. A comprehensive comparison has been given (Cooper, 1974). The points made there are summarized below.

◊ The plate heat exchanger is best for liquid-liquid duty with (flow rate)(specific heat) product nearly the same for the two fluids. Flows with dissimilar products can be applied with some decrease in overall effectiveness.

◊ For high-pressure duty (above about 300 psi), a shell-and-tube heat exchanger is preferred over a plate-type device. The basic construction of plate heat exchangers makes them unable to contain high pressures.

◊ More area per unit volume of heat transfer surface is available in the plate-type device compared to virtually all other closed-type exchangers. When this characteristic is coupled with the ease of manufacture of the plate device, the plate type usually has a lower cost than any tube-type heat exchanger. However, this comparison is made between exchangers constructed from special materials, like stainless steel. Because of the typically larger surface areas of plate heat exchangers compared to the shell-and-tube type, as well as the nature of the flow in a plate device compared to tubes, fouling applications are more easily handled in the plate exchanger.

◊ If an application does not need a heat exchanger made of a special material, a shell-and-tube device may be less expensive than a plate type. This is because a plate exchanger is normally not constructed with mild steel. The shell-and-tube heat exchanger is definitely preferred when the device can be constructed from mild steel and a close approach temperature is not required.

◊ One of the more expensive components of the plate heat exchanger is the gasket, and the materials of construction of this component will normally set the temperature limit of the heat exchanger's operation. Typically, this limit will be about 500°F (260°C).

3.2.4 Phase Change Heat Exchangers

A number of heat exchanger applications involve one or both streams undergoing a *change of phase*. While this can include freezing/thawing, in fact, the most important situations are those where boiling/condensing takes place. Some of the heat exchangers described above can be used for these applications. There may be special concern for the orientation of the heat exchanger because gravity effects are often very important in the overall heat transfer performance. For example, there is a need to remove the condensed liquid readily from a *condenser* surface so that the heat exchanger can operate more efficiently.

Devices where boiling occurs offer particular challenges. A number of factors contribute to this. For example, the *boiler* in a coal-fired steam power plant must be able to transfer the heat from the hot combustion gases to sections where preheating, boiling, and superheating are taking place. Each one of these three sections offers special design challenges (B&W, 1972; Singer, 1981). Since the design of these devices is so specialized, the interested reader is referred to the literature just noted.

On the other hand, the engineer in the chemical process industry (CPI) faces a whole range of different kinds of problems in the design and specification of *evaporators.* The CPI engineer usually designs for the separation of two or more components in a stream, while the power plant application generally has the more direct end goal of steam generation. The power plant application generally uses highly treated water whereas the CPI applications often use streams with high levels of "impurities." (This term is used loosely because one engineer's impurities might be another's end product.) As a result, continuous blowdown is usually required. Also, when a mixture of substances is used, rather than a single substance like water, the resulting mixture experiences a boiling point rise as the evaporation takes place. Finally, it is often the case in the CPI that the desired product is the residue; while in the power plant application, the goal is the manufacture of steam.

Evaporators are found commercially in a wide range of designs. These range from simple *batch-pan evaporators* that have either external steam jackets or internal heating coils to the much more complex *agitated thin-film evaporators* that can handle extremely viscous, heat-sensitive, crystallizing, and fouling materials. An excellent review of many of the types of commonly used evaporators has recently been given by Mehra (1983).

3.2.5 Air-Cooled Heat Exchangers

When heat is removed from a stream to a temperature close to ambient, air-cooled heat exchangers can be applied. While this is the most frequent use of heat exchangers where one fluid is a gas, it is certainly not the only one. The boiler, superheater, and reheat elements in a power plant are obvious, though fairly specialized, examples of these kinds of devices. Recuperators are often used where energy is recovered from a high-temperature stream. Refrigeration units almost always have some kind of device for heat transfer from a liquid and/or vapor to a gas.

The elements of design analysis for heat exchangers in a gas stream are given in many texts. Usually, it is apparent that extended surfaces (fins) are of value in decreasing the thermal resistance on the gas side of the heat exchangers. The addition of fins can increase the external area of a tube by factors up to 50 times that of a nonfinned tube (note Equation 3.1). Internally finned tubes are available also, but these are less typically applied. Extensive design and application data are given in Kays and London (1964) and Shah (1983a) and in the HVAC literature. Design techniques have even been published for programmable calculators (Shaikh, 1983).

Often something more than simply a heat exchanger will be needed for transferring heat from a liquid to a gas stream or the ambient air, and it is necessary to consider a heat transfer <u>system</u>. When removal of waste heat to the ambient is desired, several approaches may be available. Four of these options have been compared by Huber (1976), and these are shown in Figure 3.6.

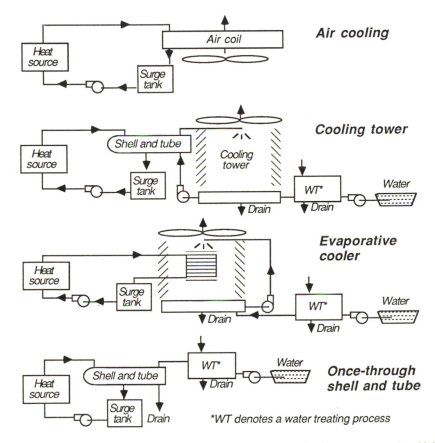

Figure 3.6 Schematic diagrams of several types of heat rejection systems, after Huber (1976).

In all cases shown in Figure 3.6, it is assumed that a liquid medium carries the heat from the source to the rejection device. Hence, a <u>pump</u>, <u>surge tank</u>, and <u>fluid circulation loop</u> is shown in each option.

In the first option shown ("air cooling"), the circulating fluid flows through a liquid/air heat exchanger denoted as an <u>air coil</u> in the figure. Use of a <u>cooling tower</u> is shown in the second option. Aspects of the first two options are shown combined in the third, called "evaporative cooler" in the figure. Finally a once-through system is shown in the third figure, where the heat is transferred to water, which is then dumped to a <u>drain</u>. In the fourth option, there may be the possibility in some instances to use the treated water directly in removing the heat from the source. Normally, however, the practice is to use the configuration shown.

Comparisons of these systems may help in many instances in determining which to apply. In terms of initial costs, the air-cooling option normally will often be the highest. This option also requires the largest land area. The once-through water system is usually the least expensive in all acquisition costs. In terms of operational costs, the water system will need more raw water and sewer capacities, and these aspects could account for sizable costs. Pollution control in any of the systems that dump water must also be considered. Electrical consumption is normally higher on any of the systems that move air. Water treatment costs must be considered on the second, third, and fourth options. Maintenance and repair costs tend to favor the dry air systems in almost every respect.

One of the major points of comparison for the systems shown in Figure 3.6 is the way in which the heat rejection duty is accomplished and the implications of this. The air system depends upon the dry-bulb temperature for its cooling capacity. Both the cooling tower and the evaporative cooler achieve primary fluid temperatures that approach the wet bulb. Of course, the once-through system may be independent of either of these in certain circumstances. If freezing conditions exist during part of the year, both the cooling tower and evaporative cooler may require special operation. The evaporative cooling option may be able to operate in the wet/dry modes, where spray water is not used in cold periods. With appropriate design, this option may perform similarly to the air-cooling option when the spray water is not used.

3.2.6 Direct-Contact Heat Exchangers

3.2.6.a General Comments

A very large number of configurations can be classified under the category of direct-contact heat exchangers. Included are applications where solids, liquids, or gases exchange heat with liquids or gases. Devices of this type range from the often applied (e.g., evaporative coolers, cooling towers, and open feedwater heaters) to forms that are much less frequently used (e.g., coolers for air compressors or heat-reclaiming devices for combustion processes).

General aspects of direct-contact heat exchangers have been discussed recently (Boehm and Kreith, 1986; Boehm and Kreith, 1987). A very large set of possibilities exists for this general category of devices. Included are very high temperature solids/gas heat transfer applications to very low temperature difference, heat transfer situations between immiscible liquids. One example of the latter type of device is shown in Figure 3.7.

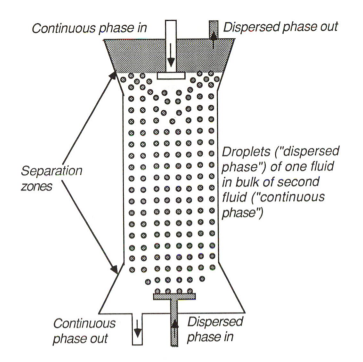

Continuous phase in   Dispersed phase out

Separation zones

Droplets ("dispersed phase") of one fluid in bulk of second fluid ("continuous phase")

Continuous phase out   Dispersed phase in

Figure 3.7 A schematic diagram of one of many varied types of direct-contact heat exchangers. This type is patterned after mass transfer columns and the fluid combinations can be liquid/liquid or liquid/gas.

There are a number of <u>advantages</u> to these types of devices. Direct-contact exchangers have the inherent characteristic that heat can be transferred between two streams without an intervening material surface. This virtually eliminates the detrimental corrosion and fouling aspects that can occur in conventional heat exchangers. Other benefits normally include lower approach or pinchpoint temperature differences as well as lower costs than those of closed-type exchangers. In summary, direct-contact heat exchange can eliminate fouling and corrosion of heat exchange surfaces, improve performance, and cut costs.

While this all sounds too good to be true (and many times it is true and very good), there are some <u>limitations</u> to keep in mind. First, since the two streams are in direct contact, they are necessarily at the same pressure during the heat transfer process. Thus, high differentials of pressure between the two streams are normally not possible. Also, it may be the case that the two streams should be such that one does not go into solution in the other or otherwise merge in a manner that makes their separation difficult. (Sometimes it is of value to have this happen, however. More on this follows.) Usually this does not cause the direct contact application to be impossible, but this requirement may result in the need for some creative engineering. Finally, and perhaps most importantly, the existing design techniques are less comprehensive and straightforward than is the corresponding situation with closed exchangers. While techniques have been established for the more commonly applied devices like cooling towers, and some relatively simple methods do exist for more general types (see Fair, 1972; Boehm and Kreith, 1987), this absence of generalized design techniques is a hindrance to widespread applications of these devices. Then, to summarize the limitations, both streams have to be at the same pressure, a moderate to high solubility of one of the fluids in the other can cause operational problems, and simplified design techniques are lacking.

An important aspect to note in passing is the role of mass transfer in direct-contact heat transfer devices. This aspect could appear on both the benefits and drawbacks list given above. In some situations, mass transfer is desired. For example, a wet cooling tower (discussed in the next section) functions better when larger amounts of water are evaporated into air. Another example, discussed in Section 3.2.6.c, where mass transfer is desired is the deaerating heater. There are situations, however, where direct-contact heat exchanger designs may benefit from the minimization of mass transfer and significant amount of mass transfer could be considered undesirable. An energy-reclaiming heat exchanger on a combustion exhaust gas stream may be one example of this.

3.2.6.b Cooling Towers, Evaporative Coolers

Cooling towers and evaporative coolers are very important devices in thermal systems for a variety of reasons. Two major distinctions can be made between these two categories of equipment. **(i)** Normally the output sought from cooling towers is cooled water, while the end result of an evaporative cooler is cooled air. **(ii)** Cooling towers are found in more types of configurations than evaporative coolers are. For example, seldom is an evaporative cooler found in a natural-draft design. (See discussion of this term that follows.) Both types of devices benefit from mass transfer that accompanies the heat transfer.

What follows will focus on cooling towers. Keep in mind that many of the comments are applicable to evaporative coolers also.

Empirically based design techniques for cooling towers have developed over the years and are now being partially replaced with more fundamentally based approaches. Typical design calculations have been outlined in the literature (Campbell,1987). A national organization oversees cooling-tower testing and certification (the Cooling Tower Institute in Houston, Texas), and this group has issued acceptance test procedures and certification standards for commercial units.

There are a number of forms of cooling towers. Generally, they can be divided into groups according to whether or not the air is forced through by fans (sometimes called *mechanical draft*) or flows by free convection (sometimes called *natural draft*). Subdivisions of the mechanical draft types are made according to whether the fan is located where the air flow comes in *(forced draft)* or where the air exits *(induced draft)*. Further subdivisions of all types of cooling towers are based on the direction of the air flow, usually being either across the water flow or counter to it. One further distinction is made according to whether or not the water and the air come in direct contact. The former are termed *wet* and are the most prevalent. The latter are denoted as *dry*. There are some types that can function in either mode. Several schematics are shown in Figures 3.8 and 3.9 to illustrate some of the types of cooling towers.

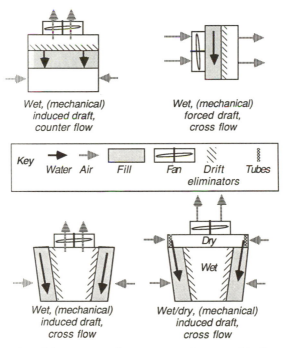

Figure 3.8 Examples of three wet cooling towers and one combination wet and dry tower. Evaporative coolers function very similarly to the wet types shown.

An excellent review of the various types of cooling towers available and their characteristics has been provided by Robertson (1980). Several aspects of operational interest have been covered by Elliott (1985).

There are relatively clear design choices between the various types of cooling towers. Although there can be exceptions, some of the considerations are noted below (Robertson, 1980).

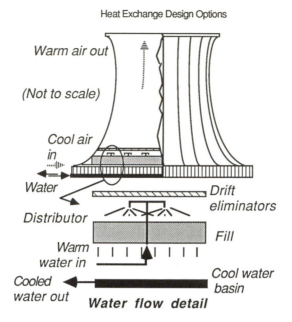

Figure 3.9 A sketch of a hyperbolic, natural-draft cooling tower.

◊ Natural-draft cooling towers are more influenced in their operation by plant and environmental conditions than are their forced-draft counterparts. This is because the natural-draft air flow rate varies with a number of environmental factors.

◊ Water pumping costs are generally less in a mechanical draft tower.

◊ Overall auxiliary power consumption is less in a natural-draft cooling tower because the decrease from not having to move air is significantly greater than the increased pumping costs.

◊ Maintenance costs are appreciably greater for mechanical draft cooling towers.

◊ More land area, site preparation, and piping costs are required for mechanical draft cooling towers.

◊ Natural-draft cooling towers usually suffer from less environmental impacts, such as ground-level fog, and recirculation problems because the moist exhaust air is rejected at higher elevations than the corresponding mechanical draft type.

◊ Capital cost comparisons reflect a clear advantage for the mechanical draft type at small installation size; but a crossover occurs as size increases, and the natural-draft type usually shows a clear advantage at larger sizes.

◊ The physical size and shape of the natural-draft type of tower may be a major factor in whether or not this type is selected.

For power plant applications, there has been a trend toward the use of *hyperbolic* towers (Elliott, 1985). A example of one of these is shown in Figure 3.9. As these devices were making their impact on the market in the 1970s, it was believed that they would be used where the following situations were applicable: low humidity and wet-bulb temperatures existed; heavy winter loads were possible; and high inlet and exit water temperatures were present. More recent applications in humid regions have also proven to be satisfactory. Elliott (1985) notes that a 10-14°F (5.5-7.8°C) approach to the summer wet-bulb temperature is not uncommon.

Dry cooling towers are simply a closed heat exchanger where the water is circulated. Air is forced across the tubes to perform the cooling. As would be expected, the approach temperature difference can be considerably higher than that for a cooling tower using direct-contact processes; but this type of cooling eliminates water use. A compromise that greatly improves performance at the cost of small water use is the dry/wet tower. See Figure 3.8. The capital costs for this type of device are higher than are the costs of either the dry or the wet towers. Mitchell and Henwood (1978) compared various cost factors for power plants using wet, dry, or dry/wet towers for the cooling systems. Some aspects of their study are presented in Table 3.4. Note that the first line of the table is indicative of the relative capital costs of the various types of cooling systems (alone) in large sizes. The numbers in the other lines show the effects of cooling system, and <u>other items needed to boost the plant capacity back to the original rating</u>, on total plant cost and produced power.

Table 3.4

Relative Cost[a] Comparisons between Power Plants
using Wet, Dry, and Dry/5% Wet Cooling Towers[b]

Item	Wet	Dry	Dry/5% Wet
Capital costs of cooling system only[c]	1	2.05	2.80
Capital costs: cooling and extra capacity[c,d]	1	3.02	2.78
Busbar energy costs[c,d]	1	1.12	1.07

a. Costs are given relative to the "wet" type of tower. **b.** Uses cost data presented by Mitchell and Henwood (1978). **c.** Applies to a 1000-MWe (net) fossil-fired power plant in the Central Valley of California. **d.** Includes only those costs affected by type and/or design of cooling system.

3.2.6.c Deaerating Heaters

Both the power industry and the chemical process industry use deaerating heaters. These are simply direct-contact heat exchangers that allow a venting of noncondensable gases. Normally, these heaters operate at atmospheric pressure, but this may not be the case in special instances.

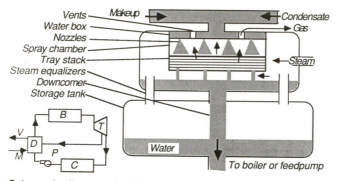

Figure 3.10 Schematic diagram of a deaerating heater patterned after Drabnis (1986). In the lower left a simplified Rankine cycle is denoted with the location of deaerating heater shown. There the following holds: B=boiler, T=turbine, C=condenser, P=condensate pump (and boiler feedpump in this situation), D=deaerating heater, M=makeup, and V=vent.

In power plant applications, the heater serves to heat condensate and make up water with bleed steam. A simplified application example is shown in Figure 3.10. Although the details of various designs differ, the one shown sprays the water onto a series of trays. Steam flows up through the trays, condenses on the water while transferring heat, and picks up noncondensables that are liberated during the heating process. This brief introduction to deaerating heaters is included to give another example where mass transfer and heat transfer occur simultaneously to good advantage in a direct-contact device.

3.2.7 Other Types of Heat Exchangers

The number of heat exchangers that will not be discussed here far outnumber those already mentioned. A classification, as well as reference to some of the important literature, of many types of heat exchangers used in the chemical process industry has been given by Lord et al., (1970). Included are double-pipe and cascade exchangers, coils and bayonet forms, spiral tube, falling-film and jacketed units, scraped-surface exchangers, thin-film heat exchangers (includes agitated film, wiped film, and centrifugal film), and froth-contact heat exchangers.

Usually, the needs dictated by new applications require special approaches. Just one of many examples of this is the use of heat exchangers in high-temperature power cycles. (See, e.g., Fraas, 1975.)

The importance of special materials of construction can have a large impact on specifics of design. For example, the use of "unusual" (from a heat exchanger point of view) materials like ceramics (e.g., Forster et al., 1980) or graphite (e.g., Hills, 1974-5) may accomplish needed heat exchanger applications with a nonconventional configuration.

3.3 THERMAL INSULATION

There are a number of purposes why thermal insulation is applied within a thermal system. Included are to:

◊ Conserve energy.

◊ Control condensation.

◊ Prevent potential injury to operating personnel by lowering surface temperatures.

◊ Maintain process temperatures.

In addition, there may be a number of nonthermal motivations for using insulation on high-temperature devices, such as decreasing emitted noise. While precise specification of insulation is normally not required in the preliminary design process, an appreciation of basic ideas of insulation application is of value.

Almost all insulations are made of solid materials. While some insulations rely on achieving a thermal conductivity nearly that of air or some other gas, the gas is usually contained within some form of solid insulation.

Insulating materials can be classified into three major categories. These are *fibrous, cellular,* and *granular.* All forms function to decrease convection and radiation. While convection is quite easily decreased to the conduction limit, the minimization of radiation normally requires some additional aspects to the basic material. For example, a basic insulating material may incorporate a low-emissivity cover to decrease radiation.

Fibrous insulations consist of thin organic or inorganic fibers manufactured in a blanket configuration. This form of insulation can be very soft and light or it can be rigid and hard.

Cellular materials are constructed so that they have small bubbles integral with a base component. Often, the manufacturing of these materials involves a "foaming" process.

Granular insulation is normally a loose-fill form, but it can be in a molded configuration also. Voids found around the particles, as well as those often found within the material, are responsible for the normally very good insulating characteristics.

Eliminate from consideration, for the moment, both extremely high temperature (above about 1100 K) and extremely low temperature (below approximately 100 K) applications. For the remaining range of applications where most thermal system designs are found, the typically encountered generic insulations are as shown in Table 3.5. Also given there are some typical values of thermal conductivity variation with temperature, the temperature application range, and density. For selected materials, the costs relative to polyisocyanurate are given for simple straight-run pipe insulation applications.

Table 3.5

Insulations for Moderate Temperature Applications[a]

Insulation Type	Thermal Conductivity W/m ˚C @-75˚C	@95˚C	@260˚C	Temperature Application Range ˚C	Density kg/m³	Relative Cost[b] for Pipe Insulation Application Diameter in.	Single Layer 2 in.	3 in.	Double Layer 4 in.	6 in.
Fibrous										
Glass	0.029	0.037	0.072	-30 to 455	85	6	1.53	2.20	2.91	4.95
						12	2.52	3.5	4.17	7.08
Mineral Wool	-	0.05	0.075	15 to 1040	128	6	1.21	1.59	2.45	3.52
						12	1.86	2.38	3.62	4.97
Cellular										
Glass	0.039	0.061	0.094	-180 to 650	136					
Polyiso-cyanurate	-	c	-	-180 to 150	32	6	1	1.53	2.22	4.08
						12	1.67	2.53	3.36	5.49
Polystyrene	0.026	d	-	-40 to 135	32					
Polyurethane	0.020	0.036	-	-75 to 100	32					
Granular										
Calcium silicate	-	0.061	0.079	15 to 815	176 to 240	6	1.45	1.92	3.06	4.46
						12	2.26	2.92	4.49	6.84
Perlite	-	0.068	0.084	50 to 815	80 to 208					

a. Modified from information given by Liss (1986). b. Installed costs of straight pipe with 0.016-in. aluminum jacket with moisture barrier (when required). Numbers given are the cost of particular insulation divided by the cost of 2 in. of polyisocyanurate insulation on a 6-in. pipe. Based upon a 1985 cost survey in the U.S. Gulf Coastal region. c. At 24˚C the thermal conductivity is approximately 0.020. d. At 24˚C the thermal conductivity is approximately 0.033.

Some variation in thermal conductivity values is noted between the various insulations shown in Table 3.5. If the variation is analyzed, it is found that there is an approximately inverse correlation between these values and the cost for a given thickness. To express this another way, there is a relationship between cost and *R* value (the thermal resistance of the insulation) such that the product is approximately constant among the insulations. Other factors that bear upon choices that are more expensive per unit *R* value involve flammability and moisture-absorption considerations. For example, the cellular insulations tend to be combustible, with glass being the single exception of those shown. On the other hand, the cellular insulations tend to absorb very little moisture, and this factor may be very important in some applications. Sometimes a less expensive insulation (on a per unit *R* basis) is eliminated because of space limitations in the installed location.

High-temperature insulations are normally made of materials that are more dense than the insulations noted in Table 3.5. As a result, the higher temperature types normally have higher values of thermal conductivity. These insulations have been found in a variety of forms over the years. *Sprayed asbestos* was used quite extensively in previous eras, but concerns about health effects of asbestos have virtually eliminated this particular material from all applications. Other forms of sprayed insulation have replaced asbestos. *Firebrick* is heavily used in high-temperature applications. This material is made by mixing an organic material with the refractory before molding and then by burning off the organic material during firing. As a result, voids permeate the brick. Another common form of high-temperature insulation consists of *ceramic fibers.* Here various fibers are mixed with binders for castable refractories. Even though plastics contain organics that are generally not considered for high-temperature applications, do not overlook them for transient heating applications. For example, one approach to the insulating of reentry vehicles from outside the earth's atmosphere is to use an *ablating* (burning) surface made of an organic material. Some of these forms can protect against temperatures in excess of 8000°C for short periods of time.

For purposes of comparison to the lower temperature insulations shown in Table 3.5, Table 3.6 shows some similar data for <u>refractories that are normally manufactured in block form</u>. As is the case with the lower temperature insulations, the density of the material can have a profound effect on the value of the thermal conductivity and the cost.

Insulations for very low temperatures offer special challenges. If the temperatures are low enough, most gases that may be present could liquify. This, of course, will minimize convection. Generally, radiation is the dominant, and sometimes virtually the only, mechanism of heat transfer. The classical approach to thermal design for low temperatures is to use evacuated regions for insulation and to apply a very low emissivity to the surfaces enclosing the vacuum space. Placement of radiation shields in the evacuated space will further decrease the heat gain.

63

Table 3.6
Selection Data for Block Form Refractories[a]

Material	Thermal Conductivity, W/m°C 260°C	538°C	816°C	1093°C	Density, kg/m³	Relative[b] Cost Installed, Flat
Mineral wool block	0.066	0.109	-	-	240	1.0
Superfex block	0.075	0.130	-	-	240	1.04
	0.090	0.111	-	-	384	1.14
Ceramic fibers						
Fiber chrome	0.072	0.180	0.187	0.272	128	3.44
Cerafelt	0.072	0.115	0.187	0.272	128	2.36
	0.072	0.101	0.153	0.215	192	3.22
Insulating firebrick	0.126	0.153	0.183	0.195	496	1.26
	0.298	0.320	0.350	0.412	816	1.42

a. Data calculated from that given by Neal and Clark (1976). b. Cost is relative to the cost for mineral wool block.

REFERENCES

ASME, 1980, ASME BOILER AND PRESSURE VESSEL CODE, American Society of Mechanical Engineers, New York.

API, -, HEAT EXCHANGERS FOR GENERAL REFINERY SERVICE, API Standard 660, American Petroleum Institute, Washington, DC.

B&W, 1972, STEAM/ITS GENERATION AND USE, The Babcock & Wilcox Company, New York.

Bellotty, J., and D. Stock, 1979, "A Numerical Design Scheme for Concentric Heat Exchangers," ASHRAE Transactions, Vol. 85, Part 2.

Boehm, R., and F. Kreith, 1986, "Direct Contact--a High Performance, Low Cost Option in Heat Exchange," Mechanical Engineering, March, pp. 78-81.

Boehm, R., and F. Kreith (Eds.), 1987, DIRECT CONTACT HEAT EXCHANGERS, Hemisphere Publishing, Washington, DC. (To appear.)

Brown, T. R., 1986, "Use These Guidelines for Quick Preliminary Selection of Heat-Exchanger Type," Chemical Engineering, February 3, pp. 107-108.

Burke, P., 1982, "Compressor Intercoolers and Aftercoolers: Predicting Off-Performance," Chemical Engineering, September 20.

Butterworth, D., 1977, "Developments in the Design of Shell-and-Tube Condensers," ASME Paper 77-WA/HT-24.

Butterworth, D., and L. Cousins, 1976, "Use of Computer Programs in Heat-Exchanger Design," Chemical Engineering, July 5, pp. 72-76.

Caciula, L., and T. Rudy, 1983, "Prediction of Plate Heat Exchanger Performance," AICHE SYMPOSIUM SERIES, HEAT TRANSFER--SEATTLE, AIChE, New York, pp. 76-89.

Campbell, J. C., 1987, "Thermal Design of Water-Cooling Towers," in DIRECT CONTACT HEAT EXCHANGERS (Boehm, R., and F. Kreith, Eds.), Hemisphere Publishing, Washington, DC. (To appear.)

Clark, D., 1974, "Plate Heat Exchanger Design and Recent Development," The Chemical Engineer, May, pp. 275-279.

Cooper, A., 1974, "Recover More Heat with Plate Heat Exchangers," The Chemical Engineer, May, pp. 280-285.

Crane, R., and R. Gregg, 1983, "Program for Evaluation of Shell-and-Tube Heat Exchangers," Chemical Engineering, July 25, pp. 76-80.

Dart, R. H., and J. Whitbeck, 1980, "Brine Heat Exchangers," in SOURCEBOOK ON THE PRODUCTION OF ELECTRICITY FROM GEOTHERMAL ENERGY (J. Kestin et al., Eds.), U. S. Department of Energy Publication, DOE/RA/4051-1, pp. 379-413.

Drabnis, A., 1986, "Selecting a Deaerating Heater to Suit Your Needs," Power, May, pp. 42-45.

Elliott, T. C., 1985, "Cooling Towers--A Special Report," Power, December, pp. S·1-S·16.

Fair, J., 1972, "Designing Direct Contact Coolers/Condensers," Chemical Engineering, June, pp. 91-100.

Fanaritis, J., and J. Bevevino, 1976, "Designing Shell-and-Tube Heat Exchangers: How to Select the Optimum Shell-and-Tube Heat Exchanger," Chemical Engineering, July 5, pp. 62-71.

Feldman, K., D. Lu, and L. Cowley, 1976, "Computer Aided Design of a Shell and Tube Heat Exchanger for Oil and Asphalt Heating," ASME paper 76-WA/HT-6.

Forster, S., et al., 1980, "Development of High Temperature Ceramic Plate-Type Heat Exchanger-and Burner-Elements," ASME/AIChE Heat Transfer Conference Paper.

Fraas, A. P., 1975, "Heat Exchangers for High Temperature Thermodynamic Cycles," ASME Paper 75-WA/HT-102.

Fraas, A. P., and M. N. Ozisik, 1965, HEAT EXCHANGER DESIGN, Wiley, New York.

Gutterman, G., 1980, "Specify the Right Heat Exchanger," Hydrocarbon Processing, April, pp. 161-163.

HEI, 1978, STANDARDS FOR STEAM SURFACE CONDENSERS, Heat Exchange Institute, Cleveland, Ohio.

Hills, D., 1974-5, "Graphite Heat Exchangers--I," Chemical Engineering, December 23, 1974, pp. 80-83. "Graphite Heat Exchangers--II," Chemical Engineering, January 20, 1975, pp. 116-119.

Huber, F. V., 1976, "Basic Principles, Advantages, and Limitations of Air-Cooled Heat Exchangers," Plant Engineering, June 24.

Karaç, S., A. Bergles, and F. Mayinger (Eds.), 1981, HEAT EXCHANGERS--THERMAL HYDRAULIC FUNDAMENTALS AND DESIGN, Hemisphere Publishing, Washington, DC.

Karaç, S., R. Shah, and A. Bergles (Eds.), 1983, LOW REYNOLDS NUMBER FLOW HEAT EXCHANGERS, Hemisphere Publishing, Washington, DC.

Kays, W., and A. London, 1964, COMPACT HEAT EXCHANGERS, SECOND

EDITION, McGraw-Hill, New York.

Kern, D. Q., 1950, PROCESS HEAT TRANSFER, McGraw-Hill, New York.

Liss, V. M., 1986, "Selecting Thermal Insulation," Chemical Engineering, May 26, pp. 103-105.

Lord, R., P. Minton, and R. Slusser, 1970. "Design of Heat Exchangers," Chemical Engineering, January 26, pp. 96-118.

Mehra, D., 1983, "Shell-and-Tube Heat Exchangers," Chemical Engineering, July 25, pp. 47-56.

Mitchell, R., and K. Henwood, 1978, "Cost of Conserving Water in Power Plant Cooling," Preprint 3346, American Society of Civil Engineers.

Neal, J., and R. Clark, 1976, "The Effective Use of Insulating Refractories," Johns-Manville Application Information, Johns-Manville, Denver, Colorado.

Pettigrew, M., L. Carlucci, P. Ko, G. Holloway, and A. Campagna, 1983, "Computer Techniques to Analyse Shell-and-Tube Heat Exchangers," ASME Paper 83-HT-61.

Robertson, R. C., 1980, "Cooling Towers," in SOURCEBOOK ON THE PRODUCTION OF ELECTRICITY FROM GEOTHERMAL ENERGY (J. Kestin et al., Eds.), U. S. Department of Energy Publication, DOE/RA/4051-1, pp. 611-644.

Shah, R., 1983a, "Compact Heat Exchanger Surface Selection, Optimization, and Computer-Aided Thermal Design," in LOW REYNOLDS NUMBER FLOW HEAT EXCHANGERS (Ed. by S. Karaç et al.), Hemisphere Publishing, Washington, DC, pp. 845-874.

Shah, R., 1983b, "Heat Exchanger Basic Design Methods," in LOW REYNOLDS NUMBER FLOW HEAT EXCHANGERS (Ed. by S. Karaç et al.), Hemisphere Publishing, Washington, DC, pp. 21-72.

Shaikh, N., 1983, "Estimate Air-Cooler Size," Chemical Engineering, December 12, pp. 65-68.

Singer, J. (Ed.), 1981, COMBUSTION POWER SYSTEMS, THIRD EDITION, Combustion Engineering Inc., Windsor, CT.

Taborek, J., and G. Hewitt (Eds.), 1982, ADVANCEMENT IN HEAT EXCHANGERS, Hemisphere Publishing, Washington, DC.

TEMA, 1978, STANDARDS OF TUBULAR EXCHANGER MANUFACTURERS ASSOCIATION. SIXTH EDITION, Tubular Exchanger Manufacturers Association, Tarrytown, New York.

PROBLEMS

Use texts on heat transfer, thermodynamics, and other topics as needed for assistance in working the following problems. Some of the problems listed here are meant to be a review of calculational procedures covered in basic courses, while others apply specifically to information given here.

3.1 Develop a computer program that will make Effectiveness-NTU calculations for either simple counterflow or simple parallel-flow shell-and-tube heat exchangers (i.e., the program, on an input cue, will make either type of calculation). Assume that overall heat transfer coefficient, U, as well as both mass-flow-rate-specific-heat products are read in as data and that one of the following two situations will always hold:

(a) The area of the heat exchanger and both incoming temperatures are given as data, and the outgoing temperatures are desired as output.

(b) One outgoing and both incoming temperatures are given as data, and the area of the heat exchanger is desired as output.

3.2 An evaporative cooler, similar to the device shown in the upper right-hand section of Figure 3.8, is used to cool air. The ambient conditions include $P = 1$ atm, $T = 40°C$, and relative humidity of 15%.

(a) What is the minimum temperature that can be achieved by the air? What will be the amount of water required (per kilogram of dry air) to accomplish this?

(b) If the air exits at 25°C and the fan flow is rated at 150 m^3/min, find the exiting relative humidity and the amount of water used in a 6-h duty.

3.3 A wet, counterflow cooling tower has an inlet water flow rate of 40 kg/s at 40°C, and the cooled water exits at 18°C. The ambient air at 20°C and 10% relative humidity flows in a direction counter to the water. If the air/water mixture leaves the tower at a dry-bulb temperature of 32°C and a wet-bulb temperature of 25°C, find the makeup water flow rate that is required. Assume that the pressure is 1 atm throughout and that there is no heat transfer, except between the air and water.

3.4 Water at 15°C is to be heated with an oil (thermal properties like engine oil) at 110°C in a shell-and-tube heat exchanger. The heat exchanger is a U-tube configuration with one shell pass and two tube passes made of 1-cm-diameter thin-walled tubes. There is approximately 1 m^2 of area. The amount of oil available is 4 kg/s. Show the resulting water exit temperature as a function of the water mass flow rate.

(a) Without making any calculations, should the oil flow through the tubes or the shell?

(b) Perform the calculations for the water flowing through the tubes.

(c) Perform the calculations for the oil flowing through the tubes.

67

3.5 It is desired to find the off-design performance of a counterflow aftercooler on an air compressor (water cooling the exhaust air). At design conditions, with the flow of air being 24.3 kg/min throughout the heat exchanger and the water flow being 30.3 l/min at its 16°C entry, the air is cooled from 121 to 38°C. At off-design, the air flow is increased to 28.2 kg/min and the entering water flow is 22.7 l/min at 27°C. If the air flow enters at 135°C, find the corresponding air exit temperature (Burke, 1982). Can this calculation be made as easily with the ε-NTU and the modified LMTD methods? If so, use both methods and compare the results. If not, use the simpler method.

3.6 Compare the various cooling options shown in Figure 3.6. Assume the following: all mass-flow-specific-heat products are the same throughout all heat exchangers; all liquid/liquid heat exchangers have an effectiveness of 75% and the value for all liquid/gas closed heat exchangers have an effectiveness of 60%; the heat source generates a 15°C temperature difference in the coolant fluid after flowing through the heat source; the dry-bulb temperature is 30°C; and the relative humidity is 20%. Make whatever further assumptions are necessary to rank the various concepts from most efficient to least efficient. Define your criteria for efficiency. List other factors specifically that should be considered in choosing between the various concepts.

3.7 Check with a vendor to determine the cost of heat exchangers with approximately the following specifications (data may not be applicable to all situations): 2 m^2 area, 0.5-cm diameter tubes, 14-bar tube pressure rating, 2-bar shell pressure rating, and temperature application range from 15 to 150°C. Do this for the following configurations:

(a) Shell and tube.

   (i) Low carbon steel tubes.

   (ii) Monel tubes.

(b) Floating head configuration of any specific type.

(c) Plate.

Where possible, compare your results to information given in Tables 3.1 and 3.2.

3.8 List all types of heat exchangers mentioned in this chapter that could be used for the following duty: heating of large flow rates of water with high flow rates of oil at temperatures between 30 and 300°C. For each of the types listed, note any further considerations you may have to evaluate before selecting that particular type.

before selecting that particular type.

3.9 Assuming that 40% of the cost of installing 1 in. of cellular glass is due to the basic cost of the material and 60% is due to labor, and further assuming that the absolute cost of the latter does not change if thicker material is used, plot a variable related to the total installed cost of the cellular glass insulation against insulation thickness, the latter ranging from 1 in. up to 12 in. thick. On the same plot show the total R value of the insulation over this same range of thicknesses. Finally, show on the same plot how the heat transfer varies through the insulated system. For this, assume that there is an external R equivalent to the value of 1 in. of this insulation and that a 100°C temperature difference exists across the whole system. Assume further that this value of external R and temperature difference do not change with insulation thickness.

3.10 Rank the following heat exchanger types for service in each of the separate situations shown below. If you cannot distinguish between two types for a given situation, or you need more information, so state. **(a)** Shell and tube. **(b)** Fixed tubesheet. **(c)** Plate. **(d)** Direct contact.

(i) Heavy fouling propensity of one fluid.

(ii) One fluid is extremely viscous.

(iii) Large ranges of temperature difference are encountered in short times.

(iv) High pressures are found in one stream.

(v) High pressures are found in both streams.

(vi) Very close approach temperatures are required.

3.11 A cooling tower is to be installed on a chemical processing plant in the driest portion of Arizona. Indicate the options you have and the information you would seek to determine which is the best type of cooling tower to install. Do not be concerned with brand names; simply focus on generic types.

CHAPTER 4

FITTING DATA AND SOLVING EQUATIONS

4.1 DATA FITTING

4.1.1 Introduction

Physical data surround us and are critically needed for the design process. This information may result from data gathering on an experiment, or it may be found in a manufacturer's product sheet, just to cite two examples. Even the data required to perform an optimal design of a complex thermal system could be almost limitless. Since the options for most designs are so numerous, the designer must have capabilities for generating mathematical formulations of physical data pertinent to the task at hand.

Depending upon the situation, the required data may be readily available within the designer's group. This is particularly true in established companies that have had a long history of design of specific items. The internal combustion engine is a typical example of this. Most heat exchanger manufacturers are another common possibility of this. They may have performance and cost data information that could be readily available for typical "new" designs.

There are instances, however, in even the most established engineering design groups when some new data might be required. Sometimes it may be a single piece of information, such as the price of 100 barrels of fuel oil delivered to Tobruk. Generally, though, the design process will require a range of information, such as the flow and power characteristics and price information on several models of blowers.

If the design is to be carried out with the aid of a computer, as assumed here, it will be desirable to reduce this data to one or more mathematical correlation(s). This is termed *curve fitting*, or, as we shall refer to it, *fitting*. While most computer systems (and many calculators!) have some kind of software that is handily used for fitting, some basic insights into this topic may be of value.

This section outlines some of the fundamental ideas of fitting. In the following subsection, the curve shapes are discussed that are represented by simple equations of one variable. A supplement to this information in the form of a brief catalog of function plots is given in Appendix C. In the third subsection given here, attention is directed to the fitting of equations when there is only enough data and when there is more than enough data to determine the parameters of the equation. The final subsection covers the fitting of data with more than one independent variable.

4.1.2 Functions of One Independent Variable, "Exact Fit"

When only one independent variable is present, then Equation 4.1 is applicable.

$$y = f(x) \tag{4.1}$$

The fitting of engineering data to functions of this form has been described by Daniel and Wood (1971) and Kolb (1982). The simplest situation is to consider the *linear* case, represented by Equation 4.2.

$$y = a_0 + a_1 x \tag{4.2}$$

Equation 4.2 has many more ramifications than might appear at first. This is due to several reasons. First, the correspondence between much of the data encountered is indeed linear. Second, if the data show a linear correspondence between the independent and dependent variables, there is a simple association of the variables that can be clearly seen. Third, there are some situations when nonlinear relationships can be cast into the linear form shown in Equation 4.2. As the old saying goes, "Be wise-- linearize." This approach to analysis can be quite helpful in many instances to simplify what could be a much more complicated analysis.

Another form of linearity is also important. This applies to the coefficients in the equation. Equation 4.2 is not only linear in x, but it is also linear in the coefficients a_0 and a_1. Because of this latter feature, the equation will be denoted here as having *linear coefficients.* Linear-coefficient equations yield to direct solution for the coefficients and to the simple application of *least-squares fitting,* which is discussed later in this section. The least-squares fitting is so important that it is of value to reduce seemingly nonlinear-coefficient equations to linear-coefficient forms wherever possible. An example of this is the simple *power function* form found in the correlation of heat transfer data.

$$y = ax^n \tag{4.3}$$

While y clearly varies in a nonlinear fashion with x, the equation can be cast in linear form by taking the logarithm[1] of both sides.

---

1 In this text "log" will be taken to be the logarithm to the base 10 and "ln" will denote the natural logarithm.

Probably the single most used function for fitting of data is the *polynomial representation.* A general form is given by Equation 4.4.

$$y = \sum_{i=0}^{n} a_i x^i \qquad (4.4)$$

While any number of terms can be applied, from a practical standpoint it is almost never the case that an order greater than four (i.e., $n > 4$) is used. In fact, most data will be fitted by no greater than a second-order form as shown in Equation 4.5. Equation 4.5 shows that although there is a nonlinear aspect between x and y that cannot be removed by a transformation, this equation is linear in its coefficients as shown.

$$y = a_0 + a_1 x + a_2 x^2 \qquad (4.5)$$

A number of functions, including the polynomial representation, are discussed and plotted in Appendix C of this text. Use this appendix for selecting appropriate types of functions to match the basic forms of data you need. In all cases, the linear or nonlinear nature of the coefficients is also mentioned.

Finally, it should be noted that sums of the various types of equations can be used to represent hard-to-fit data. In fact, many of the curve fitting programs have this capability. The problem is that intuition, which can be beneficial in selecting which functional form should fit the data, is of no value when the forms become complex. Excellent discussions of the general topic of curve fitting are given in a number of places, including works by Maron (1982) and Viswanathan (1984).

Consider now how the various equations discussed above are actually fit to the data of interest. In this discussion, attention will be given only to equations with the linear coefficient forms. The focus on this subset will not be a severe restriction. As is shown in Appendix C, a number of mathematical equations have linear coefficients that represent a large variety of curve shapes.

To solidify the idea of linear coefficients, consider the relationship shown in Equation 4.3. Taking the natural logarithm of both sides of this equation yields Equation 4.6.

$$\ln y = \ln a + n \ln x \qquad (4.6)$$

If variables are defined such that $Y \equiv \ln y$, $X \equiv \ln x$, and $A \equiv \ln a$, Equation 4.7 results.

$$Y = A + nX \qquad (4.7)$$

In general, equations of one independent variable with linear coefficients can be written in the form of Equation 4.8.

$$y = \sum_{i=1}^{n} a_i \, f_i(x) = F(x) \tag{4.8}$$

Thus, there are n unknowns in this general linear representation. To determine these unknowns, n pairs of (x_i, Y_i) data must be given, where y is used for the predicted value and Y_i denotes the data value. The curve fit then becomes a problem of finding the n values of a_i from the known information. In matrix form, this is given as

$$\begin{bmatrix} f_1(x_1) & f_2(x_1) & \cdots & f_n(x_1) \\ f_1(x_2) & f_2(x_2) & \cdots & f_n(x_2) \\ \cdots & \cdots & \cdots & \cdots \\ f_1(x_n) & f_2(x_n) & \cdots & f_n(x_n) \end{bmatrix} \begin{bmatrix} a_1 \\ a_2 \\ \cdots \\ a_n \end{bmatrix} = \begin{bmatrix} y_1 \\ y_2 \\ \cdots \\ y_n \end{bmatrix} \tag{4.9}$$

In the remainder of this text, the following matrix format will be used to represent this type of an equation in shorter notation:

$$[f]\{a\} = \{y\} \tag{4.9a}$$

Brackets ([ ]) around a boldface letter will denote a square matrix, while braces ({ }) around a boldface letter will denote a vector representation (a one-column-wide matrix). The unknown coefficients a_i are found by inverting the $[f]$ matrix.

$$\{a\} = [f]^{-1}\{y\} \tag{4.10}$$

EXAMPLE

Data are available for the cost of small electrically driven fans. If a 6-in.-diameter fan costs $13 and a 12-in.-diameter fan costs $18, find a curve fit for predicting the price of these fans. It is known from the study of fans that the cost of this model varies with the diameter of the fan in a form similar to Equation 4.3.

The linear equations that must be solved use the alternate form similar to that shown in Equation 4.4, but we will arbitrarily use the base 10 logarithm here. Using the given data, the numbers are as follows:

$$log\ 13 = 1.114 = log\ a + n\ log\ 6 = a + 0.778\ n$$

$$log\ 18 = 1.255 = log\ a + n\ log\ 12 = a + 1.079\ n$$

In matrix form, this is given as

$$\begin{bmatrix} 1.0 & 0.778 \\ 1.0 & 1.079 \end{bmatrix} \begin{bmatrix} a \\ n \end{bmatrix} = \begin{bmatrix} 1.114 \\ 1.255 \end{bmatrix}$$

This system is easily cast into a triangular form by subtracting the first equation from the second (this approach is discussed in greater detail when the solution of systems of equations is treated later in this chapter). It is then found that $n \approx 0.468$, $log\ a \approx 0.750$, and $a \approx 5.623$. Or

$$\$ = 5.623d^{\,0.468} \qquad \text{(answer)}$$

Checking this result, find that at $d = 6$, the curve fit yields a cost of $13.006, while at $d = 12$, the curve fit yields a cost of $18.000.

The general approach outlined above is used often. When this is done, keep in mind what situation exists. <u>The number of data pairs exactly equals the number of constants to be determined in the equation.</u> For the curve fit to be accurate, the data pairs have to have a low uncertainty and be fit by a curve that adequately describes the data over the range of interest. Thus, the analyst must have insight into the proper form of the mathematical relationship of the data at hand, and the data must be typical and/or precise. This is denoted as the *exact-fit* case.

4.1.3 Least-Squares Technique, "Best Fit"

Many times quite a different situation than that described in the "exact-fit" case is encountered. There may be considerably more data pairs available than the number of free parameters in the equation that is appropriate. In this circumstance, it may be desired to achieve the "best fit" that the data will allow. This is where the classical concept of *least-squares fit* comes into play.

The idea behind the concept of least squares is very simple. This technique is a way of choosing the parameters in a curve fit such that the differences between the predicted values and the initial data are minimized. However, the criterion is not simply that the difference is made as small as possible. If it were, then a line halfway between two data points at the same value of the independent parameter would yield a zero error. (This is the case since the difference between the prediction and the high point would just exactly balance the negative value between the prediction and the low point. An indication of zero error would result.) Instead, the concept of least-squares fit uses the square of the differences between the prediction and the actual data and minimizes these values.

To demonstrate the ideas involved in a least-squares fit, the concept of a *residual* is important. The residual is the difference between the prediction and the data point at the same value of the independent variable. At the same value of the independent variable, denote the predicted values

74

by the notation y_i and the data values as the symbol Y_i. Thus, the residual is the quantity given by the symbol r_i in Equation 4.11.

$$r_i \equiv Y_i - y_i \qquad (4.11)$$

The least-squares fit then requires that the sum of the squares of the residuals be minimized to achieve the *best fit*. Since any number of parameters must be determined in a given curve fitting equation (usually, though, the number will be between 1 and 4), the sum of the squares of the residuals is minimized with respect to each of the parameters. This is best visualized by examining a specific case.

Consider a second-order polynomial as shown in Equation 4.5. At each x_i value, there is a predicted value, y_i, and an actual value, Y_i. The difference is the residual, as has been noted already. Squaring all residuals and taking their sum, we have

$$\sigma \equiv \sum_{i=1}^{n} r_i^2 = \sum_{i=1}^{n} (Y_i - a_0 - a_1 x_i - a_2 x_i^2)^2 \qquad (4.12)$$

The parameters to be determined are a_0, a_1, and a_2, and these will be set to make the sum a minimum value. This is accomplished by taking the derivative of Equation 4.12 with respect to the three parameters.

$$\frac{\partial(\sigma)}{\partial a_0} = 0 = \sum_{i=1}^{n} 2(Y_i - a_0 - a_1 x_i - a_2 x_i^2)(-1) \qquad (4.13a)$$

$$\frac{\partial(\sigma)}{\partial a_1} = 0 = \sum_{i=1}^{n} 2(Y_i - a_0 - a_1 x_i - a_2 x_i^2)(-x_i) \qquad (4.13b)$$

$$\frac{\partial(\sigma)}{\partial a_2} = 0 = \sum_{i=1}^{n} 2(Y_i - a_0 - a_1 x_i - a_2 x_i^2)(-x_i^2) \qquad (4.13c)$$

If the sum in each portion of Equation 4.13 is expanded, a series of three linear equations in the three unknowns (a_0, a_1, and a_3) results, as is shown in Equation 4.14. This system of equations is easily solved to yield the desired coefficients. Note that the (x_i, Y_i) pairs are the data values given in the problem.

$$\sum_{i=1}^{n} Y_i = a_0 n + a_1 \sum_{i=1}^{n} x_i + a_2 \sum_{i=1}^{n} x_i^2 \qquad (4.14a)$$

75

$$\sum_{i=1}^{n} x_i \, Y_i = a_0 \sum_{i=1}^{n} x_i + a_1 \sum_{i=1}^{n} x_i^2 + a_2 \sum_{i=1}^{n} x_i^3 \qquad (4.14b)$$

$$\sum_{i=1}^{n} x_i^2 \, Y_i = a_0 \sum_{i=1}^{n} x_i^2 + a_1 \sum_{i=1}^{n} x_i^3 + a_2 \sum_{i=1}^{n} x_i^4 \qquad (4.14c)$$

Before ending this subsection, two observations should be made. First, keep in mind that although the constants found as outlined above will minimize the square of the prediction error, an appropriate form of the equation must be known or the fit could be meaningless. A good check on this aspect is to plot the residuals as a function of the independent variable. If the fit is good, the residuals will show no particular trend. Examples of curve fits **(a)** requiring a better equation to describe the data and **(b)** not requiring a better equation are shown in Figure 4.1. In the left-hand portion of the figure, the residuals show a correlation with x, which indicates that another fit might match the data better. The right-hand portion of the figure shows no particular correlation between the residuals and the independent parameter. The magnitudes of the residuals overall in the right-hand portion of the figure indicate the consistency of the data. If the residuals were all very small, the curve fit would be very good.

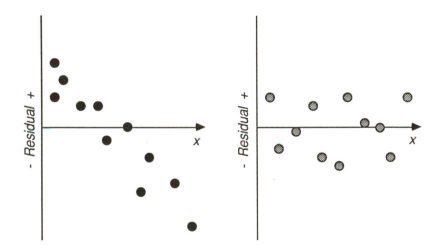

Figure 4.1 Plotting the residuals of a given curve fit against the independent variable may show whether or not the equation chosen is appropriate. The plot on the left shows a correlation of the residual with x, indicating another form of the curve fitting equation could improve the resulting prediction. Randomly distributed residuals as shown on the right usually mean the form of the chosen equation is appropriate.

Second, the emphasis given earlier in this chapter to curve fitting equations with linear coefficients should again be noted. This characteristic allows the direct solution for least squares shown in this section. Equations with nonlinear coefficients can be fit with more effort

(Daniel and Wood, 1971; Norris, 1981). Many computer systems have software that will perform these types of operations. This topic will not be covered here.

4.1.4 Two or More Independent Variables

When curve fits with more than one independent variable are needed, the corresponding procedures are more complicated than the ones used in subsections 4.1.3 and 4.1.4. Complete books have been written on this topic. (See, e.g., Daniel and Wood, 1971.) In general, though, the basic ideas for accomplishing fits to multivariable functions are related to the concepts already discussed for functions of a single variable. Depending upon the data available, either an "exact-fit" approach or a "best-fit" approach may be of value.

Consider first an exact-fit situation shown in Figure 4.2. Here a function $z = f(x,y)$ is plotted on three separate graphs. It is assumed that the data specified are limited and accurate (or all that are available), so that the data can be fit with an equation that has the same number of parameters as there are data points. In the case shown in the figure, it is assumed that a total of nine data values are given and that one parameter is held constant while a second is given for three points.

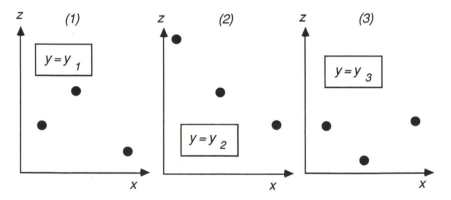

Figure 4.2 A function of two independent variables plotted parametrically. It is assumed that nine data points are given, with the values of z occurring at the same values of x (i.e., x_1, x_2, x_3) in each case.

For an "exact-fit" case when there are two or more independent variables, start the same way as the single-variable case: plot the data. Assume that this has been done as shown in Figure 4.2. Next, the form of the equation to be applied must be chosen. If something about the physics or some other basic understanding of the data can help in deciding the form to use, by all means invoke that form. Otherwise, the situation is one of choosing an equation that should fit the data over the range given. Examination of the three curves in Figure 4.2 indicates that each shape is quite different. While different forms of functions might be chosen to fit each of them, an all-purpose form such as a polynomial should be used. Then each of the three correlations should have three parameters that can be determined from the data.

For the purposes of being specific here, a polynomial is chosen to accomplish the fit. Using a simultaneous-solution approach as was described in Section 4.1.3, a curve fit is found for each of the three portions. For example, the first of these could be represented as

$$z(x, y_1) = a_1(y_1) + b_1(y_1) \, x + c_1(y_1) \, x^2, \qquad y = y_1 \qquad (4.15)$$

When all three representations like that shown in Equation 4.15 are found, the final step is to fit the functions of y_1, y_2, and y_3. This is precisely analogous to the situation shown in Equation 4.8. The specific equations that must be fit in the second step include the following three:

$$a = a_1 + b_1 \, y + c_1 \, y^2$$

$$b = a_2 + b_2 \, y + c_2 \, y^2 \qquad (4.16)$$

$$c = a_3 + b_3 \, y + c_3 \, y^2$$

Similar equations for a_2, a_3, b_2, b_3, c_2, and c_3 are determined in the second set of solutions. [Keep in mind that when the equation shown above for a is evaluated at y_1, $a_1(y_1)$ results.] The second solution set simply casts the originally determined nine constants (from equations similar to 4.15) into another set of nine constants (from equations similar to 4.16). Other ways of fitting to functions of more than one variable are essentially formalisms of the technique just outlined.

The "best-fit" analysis follows in a manner similar to that outlined for a single independent variable in Section 4.1.3. Considering a function of two variables, the generalized formulation can be written as shown in Equation 4.17. More variables than two can be handled by generalization.

$$z(x, y) = a_o + a_1 \, f_1(x, y) + a_2 \, f_2(x, y) \qquad (4.17)$$

There is no restriction on the arbitrary functions except that they cannot include undetermined constants. In other words, any constants must be in the form of linear coefficients already discussed at the beginning of this section. Thus, a permissible functional form could be the following:

$$z(x, y) = a_o + a_1 \, x \, \sin y + a_2 \, y^2 \, \ln x \qquad (4.18)$$

The multivariable situation becomes considerably more complex than its single-variable counterpart. In general, factors such as the possible dependence between the independent variables becomes of concern. The spacing of the independent data, while important in the single-variable situation, becomes even more critical here.

Readers who desire more information on this topic are referred to the definitive text by Daniel and Wood (1971).

4.1.5 Interpolation

An important aspect of curve fitting is interpolation. Often data may be available in tabular form and used that way, without the application of a complete empirical fitting equation. When this happens, a means of *interpolation* is usually needed. *Interpolation* is a means of inferring data at points that fall between given data on the number scale. What kinds of techniques are available to accomplish this?

A variety of techniques can be used for interpolation. Most texts on numerical methods contain a great deal of theory on these techniques. Included are forward difference, backward difference, and central difference formulations of various types attributed to such mathematical greats as Newton, Taylor, Stirling, Bessel, Lagrange, and Everett. Several of the approaches are similar to curve fitting concepts discussed earlier in this chapter. A limited number of techniques are dealt with here in the interpolation context.

First, it is of value to state formally what is involved in interpolation. If there is a function $F(x)$, and values are known for it at distinct points, say at x_i, then what are estimated values for the function between the given values?

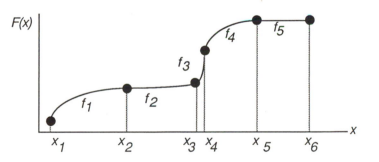

Figure 4.3 An example of a function $F(x)$. Assume that discrete pairs of (x,F) values are given at various locations x_i. It is desired to find approximating functions f_i to the spans of the actual function F.

Consider the function shown in Figure 4.3. In specific, it is desired to find *approximating functions*, denoted here as $f_i(x)$, to the actual function $F(x)$, so that values of the actual function can be estimated between any two given points, say x_i and x_{i+1}. Denote the region of the curve between any two points i and $i+1$ as the *span* of the curve in that region. Of course, it is desirable to have the approximation function be as

79

close to the actual function as possible using only the discrete values of the function. Hence, knowing the discrete points, what is an approximating relationship for the spanning regions? (A related problem is the need to estimate extreme values at either end of the given data sequence. For the sake of brevity, the latter problem will not be addressed here.)

Perhaps the most familiar of all interpolation techniques is the one that is often denoted by the name *linear interpolation*. Few designers have not enountered this concept before. In this approach, the points are assumed to be connected with straight lines. See Figure 4.4. The approximating function for the ith span is given by Equation 4.19.

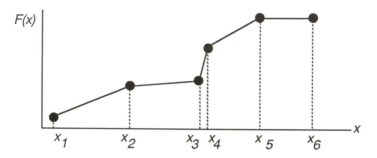

Figure 4.4 The case where given data are spanned with straight lines is the well-known linear interpolation method.

$$f_i(x) = \frac{\lceil x-x_{i+1} \rceil}{\lfloor x_i-x_{i+1} \rfloor} F(x_i) + \frac{\lceil x-x_i \rceil}{\lfloor x_{i+1}-x_i \rfloor} F(x_{i+1}) \quad (for\ x_i \le x \le x_{i+1}) \qquad (4.19)$$

For this situation, the value of the function is preserved across the given point. Or:

$$f_{i-1}(x_i) = f_i(x_i) \qquad (4.20)$$

Intuition, mathematics, and Figure 4.4 can all show, however, that the same slope is not preserved on each side of the given points. To allow a better fit, it is necessary to use a more complicated approximating function.

The linear relationship is a simple form of the general polynomial interpolation (Maron, 1982). An nth-degree polynomial will have n unknowns that can be used to match the value of the function and the $(n-1)$st derivatives of the function across the point. In practice, it is important that a compromise is made between ease of use and goodness of fit. Third-order polynomials are an example of a very good compromise. The application of these, called cubic splines, is now discussed.

Cubic splines are often used in interpolation applications. Typically, the form of these interpolating functions for each span is taken as follows:

$$f(x) = a + b\,x + c\,x^2 + d\,x^3 \qquad (4.21)$$

Not only does this approach allow the values of the function to be made the same across the given data points, as shown in Equation 4.20, but the first and second derivatives of the functions are made the same across each point. That is

$$f_{i-1}{}'(x_i) = f_i{}'(x_i) \qquad (4.22)$$

$$\text{for } i=1 \text{ to } n-1$$

$$f_{i-1}{}''(x_i) = f_i{}''(x_i) \qquad (4.23)$$

In addition, there is normally a stipulation that the second derivatives of the span functions vanish as the two endpoints are approached. From Equation 4.21, it can be seen that the second derivative of the function is linear. A sketch of two spans of a possible set of cubic splines is shown in Figure 4.5. The value and slope of both f and f' are preserved between the two spans (across the data point i). Only the value (not the slope) of f'' is the same.

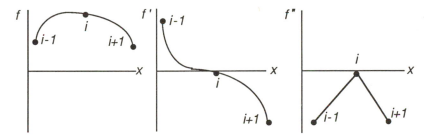

Figure 4.5 A sketch of the variation of the approximating function, $f(x)$, and its first two derivatives, $f'(x)$ and $f''(x)$, for two spans of data. The assumed linear behavior of f'' is shown.

Taking the second derivatives of the cubic spline functions to be linearly interpolated (Hang, 1976; Maron, 1982), a form like Equation 4.19 in terms of second derivatives is

$$f''(x) = \left[\frac{x_{i+1} - x}{x_{i+1} - x_i}\right] f''(x_i) + \left[\frac{x - x_i}{x_{i+1} - x_i}\right] f''(x_{i+1}) \qquad (4.24)$$

In this equation, all of the terms shown as f'' are unknowns. Integrating this equation twice, and applying the continuity of values and slopes, as described above, a system of equations like that shown in Equations 4.25 results. Here the notation of $f''_i \equiv f''(x_i)$ is used. The Greek symbols represent groups of constants. Details of the development are left to the reader. (See the Problems section.)

$$\beta_2 f_2'' + \gamma_2 f_3'' = \delta_2$$

$$\alpha_3 f_2'' + \beta_3 f_3'' + \gamma_3 f_4'' = \delta_3$$

$$..... \quad ..... \quad ..... \quad ...... \quad ..... \qquad (4.25)$$

$$\alpha_{n-2} f_{n-3}'' + \beta_{n-2} f_{n-2}'' + \gamma_{n-2} f_{n-1}'' = \delta_{n-2}$$

$$\alpha_{n-1} f_{n-2}'' + \beta_{n-1} f_{n-1}'' = \delta_{n-1}$$

Once the f_1'' through f_n'' are found (remember that f_1'' and f_n'' are assumed to be zero), then the solutions for the approximating functions are easily determined.

It has been discussed in the literature (e.g., see Hang, 1976) that inaccuracies can arise from the use of cubic splines that can be lessened with the use of higher order polynomials. Normally, the additional complexity that is required does not result in that much better of a fit. Splines are discussed in a number of sources in the numerical analysis literature, and the interested reader should have no trouble finding a great deal of additional information.

4.1.6 Relationships from Physical Concepts

As this section ends, an important point needs to be noted. Many times some good physical basis can be used in determining relationships between key variables in a given problem. Whenever facing a curve fit situation, be sure to keep in mind any physics that may be at play. Use that information to build better approximations.

For example, if relationships are to be found between variables in any kind of heat exchange element, fluid flow device, or thermodynamic process, try first to determine the theoretically based relationship(s). Sometimes the complete relationship(s) can be found directly. At other times, some empirically based factor enters into the data and modifies the theoretical performance. For the latter situation, use the known performance relationships from their theoretical basis and then determine a functional relation for the empirical factor. An example of this is the efficiency of a pump, compressor, fan, or turbine. The performance of these devices is easily described theoretically and modified with a curve fit efficiency.

This topic is not explored in any depth here. The reason for this is that the basic texts in heat transfer, thermodynamics, fluid flow, and other engineering fundamentals deal with the modeling of physical processes at great length. Keep the concepts covered there clearly in mind during your work with system simulations.

4.2 SOLUTION OF ALGEBRAIC EQUATIONS

4.2.1   Introduction

Other sections of this text deal with how to set up equations to model thermal systems. Everything from thermodynamics to costs are addressed in the model building. At some point, a number of equations must be solved. Depending upon the type of model developed, the number of equations could range from one to a few thousand. The equation(s) may be linear or nonlinear in the independent variable(s). There is a possibility that the equations could be of differential form, particularly if the transient response of a system is modeled, but this situation will not be addressed here. Hence, the numerical solution of algebraic equations will be of primary concern. Even with this restriction in topic, the field is very broad.

Often the solutions that are needed are those for <u>systems of linear algebraic equations</u> and those for <u>single and systems of nonlinear equations</u>. As will be shown, <u>the solution for systems of nonlinear equations requires the solution for systems of linear equations</u>, so the latter topic is doubly important.

When solving systems of linear algebraic equations, several characteristics about the systems are often present. First, the systems can be very large. There may be several unknowns for each block in a flowsheet, so not very many blocks are required to result in a large equation system that must be solved. Second, the systems are often *sparse.* This means that only a few of the variables appear in each of the equations. Using conventional solution techniques can be very inefficient. The focus here will be on aspects that relate to both of these facts.

Finally, it should be noted that more texts and other literature, as well as existing software, are available to the reader on this topic than any other topic covered in this book. This means that only limited, introductory discussions are realistic here. It also means that additional information and software are abundant and readily accessible. Use the material here as an introduction and continue your work by referring to the numerous other resources available.

4.2.2   Solution of Nonlinear Equations

Consider the solution of a single equation as shown in Equation 4.26, which is assumed to be nonlinear. In other words, find the root(s) of this equation.

$$F(x) = y \quad for \quad y = 0 \qquad (4.26)$$

Assume that there is only one root in the region of interest. Solutions of this type arise in many situations. An example is the determination of the operating point for the fluid flow loop consisting of a pump and pressure drop in a piping circuit. In this case, the pressure drop in the piping would be given as a function of mass flow in the form $\Delta P_1 = f_1(m)$, and this would probably be a monotonically increasing, nonlinear function that starts at the orgin of a ΔP versus flow rate plot. The pump characteristic curve would be given as $\Delta P_2 = f_2(m)$, with the sign of ΔP_2 being opposite of ΔP_1. This second function would be nonlinear also, starting from a maximum value at no flow and possibly decreasing to zero at some value of flow. The resulting equation in the form of Equation 4.26 that must be solved for m is $\Delta P_1 + \Delta P_2 = f_1(m) + f_2(m) \equiv F(m) = 0$.

To see how Equation 4.26 might yield to a numerical solution, consider a series expansion of the function around the root, defined to be x_0. This is given in Equation 4.27.

$$F(x) = F(x_0) + F'(x_0) (x - x_0) + F''(x_0) (x - x_0)^2 / 2!$$

$$+ F(x_0)''' (x - x_0)^3 / 3! \ldots \tag{4.27}$$

By hypothesis, $F(x_0) = 0$. At points sufficiently close to x_0, terms above the linear one in the expansion can be neglected, and $F'(x_0) \approx F'(x)$. With these observations, Equation 4.27 reduces to Equation 4.28a, which in turn is cast into Equation 4.28b.

$$F(x) \approx F'(x) (x - x_0) \tag{4.28a}$$

or

$$x_0 \approx x - F(x) / F'(x) \tag{4.28b}$$

In general,

$$x^{n+1} = x^n - F(x^n) / F'(x^n) \tag{4.28c}$$

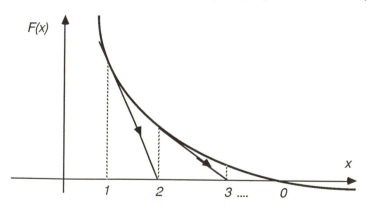

Figure 4.6 Schematic of the convergence to the root (denoted here as 0) of a function with the use of Newton's method.

This is the basis of *Newton's method,* often called the *Newton-Raphson method,* for the solution of nonlinear equations. Examination of Equation 4.28c shows that an estimate of the root x_o can be found from information about the function at an arbitrary point x. The approach is shown schematically in Figure 4.6.

While Newton's method converges very quickly to the root in most cases, there are situations when the curvature of the function will cause a diverging solution (no convergence.) An example of a case where this could be a problem is the single-root function shown in Figure 4.7. If a trial value of the x is taken larger than **xcrit**, then the calculated slopes will lead the numerical technique to seek the root at the infinite asymptote. If, on the other hand, a trial value less than **xcrit** is used, the single root will be found directly.

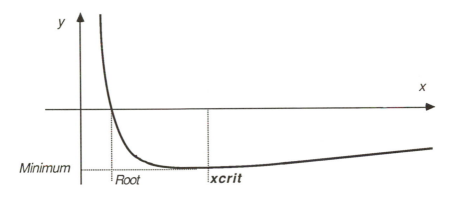

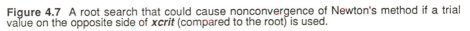

Figure 4.7 A root search that could cause nonconvergence of Newton's method if a trial value on the opposite side of **xcrit** (compared to the root) is used.

Of concern is the situation where there may be several roots, and a specific set of those roots is desired. These and other important topics about the limitations of Newton's method are addressed in a number of books. (See, e.g., Ralston, 1965; Wait, 1979; Sargent, 1981.)

Newton's method is easily extended to systems of two or more nonlinear equations. To illustrate how this is done, consider the two equations shown in Equations 4.29.

$$z_1 = f_1(x, y) \qquad\qquad (4.29a)$$

$$z_2 = f_2(x, y) \qquad\qquad (4.29b)$$

The procedure follows analogously to that used for the single-variable, single-equation example above. Expanding the function in a series about the roots, x_o and y_o, the results shown in Equations 4.30 are found. Graphical representation of the two equations is more complicated than the corresponding situation for the single equation and will not be shown here.

85

$$f_1(x,y) = f_1(x_0,y_0) + \frac{\partial f_1(x_0,y_0)}{\partial x}(x-x_0) + \frac{\partial f_1(x_0,y_0)}{\partial y}(y-y_0) + \ldots \tag{4.30a}$$

$$f_2(x,y) = f_2(x_0,y_0) + \frac{\partial f_2(x_0,y_0)}{\partial x}(x-x_0) + \frac{\partial f_2(x_0,y_0)}{\partial y}(y-y_0) + \ldots \tag{4.30b}$$

As in the single-equation situation, the series is truncated to the first-derivative terms, the fact that $f_1(x_0,y_0) = f_2(x_0,y_0) = 0$ is used, and the derivatives are assumed to have nearly the same values at the general point (x,y) as they do at the root (x_0,y_0). The system then reduces to an n x n set of linear algebraic equations (it is a 2 x 2 system here) that must be solved at each set of trial values (x,y). For the specific situation shown in Equations 4.30, the matrix system that must be solved on each iteration is shown in Equation 4.31. In all cases, the function and the derivatives are evaluated at (x,y), and the matrix is solved for the estimated correction to the roots $(x-x_0)$ and $(y-y_0)$. Note that the initial problem of solving a system of <u>nonlinear equations</u> reduces to the iterative solution of a system of <u>linear equations</u>. The latter are treated in the next section of this chapter.

$$\begin{bmatrix} \partial f_1/\partial x & \partial f_1/\partial y \\ \partial f_2/\partial x & \partial f_2/\partial y \end{bmatrix} \begin{bmatrix} (x-x_0) \\ (y-y_0) \end{bmatrix} = \begin{bmatrix} f_1 \\ f_2 \end{bmatrix} \tag{4.31}$$

In shorthand form, Equations 4.31 can be written as Equation 4.32.

$$\boldsymbol{J} \, \Delta = \boldsymbol{f} \tag{4.32}$$

The n x n matrix $\boldsymbol{J}$ is known as the *Jacobian.* The characteristics of the Jacobian play an important role in the efficient solution of large systems. (Sargent, 1981). As before, the notation is used that a capitalized boldface letter denotes an n x n matrix, while a lowercase boldface letter denotes an n-component vector.

4.2.3 Solution of Systems of Linear Algebraic Equations (Dense)

Systems of linear algebraic equations arise frequently in the modeling of thermal systems. Earlier it was shown that there are several instances where a system of linear equations could arise from a curve fitting work. Virtually any flowsheet modeling will involve the solution of linear equations. As was just discussed (Section 4.2.2), the solution of any system of nonlinear equations will reduce to the repeated solution of a system of linear equations. Hence, a variety of sources of linear equations is encountered in the design process.

Computer routines for solutions to linear systems of algebraic equations are very prevalent. It is hard to imagine someone not having access to this type of tool. For this reason, the coverage here will not focus so much on the implementation of solutions but rather on some of the key ideas behind modern techniques. In this section, general concepts are addressed, with particular emphasis on *dense* systems. This name denotes equation systems where essentially all terms are present. The final section of the chapter will address some ideas particularly relevant to *sparse* systems that have many terms missing. Sparse systems occur often in practice, and these are discussed in the next section.

Everyone has been introduced to the use of Cramer's rule for the solution of simple linear equation systems. Unfortunately, this approach is not of much value when the system is larger than 3 x 3. Since most systems encountered in the thermal system modeling are often at least an order of magnitude larger than that, some other approaches have to be used.

First consider some terminology (Forsythe and Moler, 1967). A matrix A is said to be *symmetric* if $a_{ij} = a_{ji}$ and *diagonal* if the following holds:

$$a_{i,j} \quad \begin{matrix} = 0 \text{ for } i \neq j \\ \neq 0 \text{ for } i = j \end{matrix}$$

A diagonal matrix will be denoted here as D.

Diagonally dominant denotes the following characteristic:

$$|a_{i,i}| \geq \sum_{j \neq i} |a_{i,j}|$$

for all *i*, with the inequality holding for at least one *i*. A is said to be a *band matrix* if

$$a_{i,j} = 0 \text{ for } |i - j| > m$$

The matrix A is said to be *reducible* if there exists a permutation of rows and a (usually another) permutation of columns such that a rearrangement of A yields

$$\begin{bmatrix} E & O \\ F & G \end{bmatrix}$$

where E and G are square submatrices and O is a *null* (or *zero*) submatrix.

The basic problem that is faced is the solution of a system of linear algebraic equations. In matrix form, the basic system is given by Equation 4.33.

$$A \ x = b \tag{4.33}$$

Mathematically, the solution is represented by finding the inverse of the **A** matrix, denoted by **A**$^{-1}$. This is shown in Equation 4.34.

$$x = A^{-1} b \qquad (4.34)$$

There are several ways that the solution to Equation 4.34 can be carried out numerically that do not involve actually generating an inverse matrix. The most widely applied technique is called *Gaussian elimination*. Before this approach is demonstrated, consider the system of equations shown here as Equations 4.35.

$$\begin{aligned}
10\,x_3 &= 21 \\
-4\,x_2 + 14\,x_3 &= -2 \\
3\,x_1 + 8\,x_2 - 19\,x_3 &= 0
\end{aligned} \qquad (4.35)$$

Although this is a 3 x 3 system, it is a very special one. The first equation can be solved directly for x_3. This result can be substituted into the second equation, which allows the direct solution for x_2. Then the values for both x_2 and x_3 are used in the third equation to find the value for x_1. A solution is found in a very straightforward manner! The form of the corresponding coefficient matrix for the system shown in Equations 4.35 is called *triangular*. A triangular form usually is in reference to the orientation about the major diagonal (the terms $a_{i,i}$). Two important applications are the *lower-triangular form*, which is represented by a coefficient matrix, as shown below:

$$\begin{bmatrix}
a_{1,1} & 0 & \cdots & & 0 \\
a_{2,1} & a_{2,2} & 0 & \cdots & 0 \\
& & \cdots\cdots\cdots & & \\
a_{n-1,1} & a_{n-1,2} & \cdots & & 0 \\
a_{n,1} & a_{n,2} & \cdots & & a_{n,n}
\end{bmatrix} \equiv L$$

or the *upper-triangular form*, as shown below:

$$\begin{bmatrix}
a_{1,1} & a_{1,2} & \cdots & & a_{1,n} \\
0 & a_{2,2} & \cdots & & a_{2,n} \\
& & \cdots\cdots\cdots & & \\
0 & 0 & \cdots & & a_{n-1,n} \\
0 & 0 & \cdots & 0 & a_{n,n}
\end{bmatrix} \equiv U$$

Note that the coefficient matrix for Equations 4.35 is of the upper-triangular form, as can be seen by rearrangement. Special cases of **U** and

L, where only values of unity are found on the diagonal, are denoted here as U^1 and L^1.

The fact is that virtually any set of linear algebraic equations can be manipulated into the form of Equations 4.35. This is the basis of Gaussian elimination. Although there is a number of variants on this general approach (Stewart, 1973; Johnson, 1982), only the classical technique will be dealt with here. Consider Equations 4.36, where a solution for x_1, x_2, and x_3 is desired.

$$-x_1 + x_2 + x_3 = 6$$
$$8x_1 + 2x_2 - x_3 = -7 \qquad (4.36)$$
$$7x_1 + 6x_2 + 4x_3 = 17$$

It is instructive and convenient to deal with the *augmented matrix* (defined as the constant vector **b** with the coefficient matrix **A**), as shown below:

$$\begin{bmatrix} -1 & 1 & 1 & 6 \\ 8 & 2 & -1 & -7 \\ 7 & 6 & 4 & 17 \end{bmatrix}$$

A constant times one equation can be added to another equation in the same linear system without changing the solutions to the set. Here take 8 times the first line and add it to the second line. In addition, take 7 times the first line and add it to the third line. The result is given below:

$$\begin{bmatrix} -1 & 1 & 1 & 6 \\ 0 & 10 & 7 & 41 \\ 0 & 13 & 11 & 59 \end{bmatrix}$$

Now add -1.3 times the second line to the third line. The following is the result:

$$\begin{bmatrix} -1 & 1 & 1 & 6 \\ 0 & 10 & 7 & 41 \\ 0 & 0 & 1.9 & 5.7 \end{bmatrix}$$

This is the same form as was dealt with in Equations 4.35 and the solution $\{-1\ 2\ 3\}^T$ follows in a similar manner. Use of the notation $\{\}^T$ indicates the *transpose,* where rows become columns and vice versa.

The Gaussian elimination procedure is easily generalized. The coefficients resulting from the first set of eliminations are given by

$$a_{ij}^{(1)} = a_{ij} - \frac{a_{i1}}{a_{11}} a_{1j} \qquad \begin{array}{l} i = 2, \ldots, n \\[1em] j = 2, \ldots, n+1 \end{array}$$

After k eliminations,

$$a_{ij}^{(k)} = a_{ij}^{(k-1)} - \frac{a_{ik}^{(k-1)}}{a_{kk}^{(k-1)}} a_{kj}^{(k-1)} \qquad \begin{array}{l} k = 1, \ldots, n-1 \\ i = k+1, \ldots, n \\ j = k+1, \ldots, n+1 \end{array} \qquad (4.37)$$

Here the notation $a^{(i)}$ is used to denote the results of the ith elimination, $a_{j,n+1}$ is taken as the component usually named b_j, and $a_{ij}^{(0)}$ is simply a_{ij}. It is obvious that a_{kk} cannot be zero.

Since it is easily seen that a_{kk} cannot be zero, as was noted above, then the effects of a_{kk} becoming very small should also be contemplated. Ideally, a small a_{kk} will not have any adverse effect. A problem is always present in practice because computing machines have some limit on the number of significant digits they tally. It should be obvious that a very small value of a_{kk} could cause the second term on the right-hand side of Equation 4.37 to make the first term of that equation have no effect on the overall computation.

As the number of significant digits that can be handled by the computer goes down, the need for *pivoting*, which is always desirable, becomes more important. The need for this is to preserve diagonal dominance, as defined earlier in this section. Pivoting is accomplished by bringing to the diagonal the largest element in a given set of eliminations. Adjusting the order of equations does not affect the true solution in any way, but this simple technique could affect the accuracy of the calculation of the solution profoundly. Often, though, the simple reordering of equations may not be sufficient to bring the largest coefficient to the diagonal.

Other techniques can be used to make sure that diagonal dominance is present in a set of eliminations. Clearly, the multiplication of a given equation (and thus multiplying a given row in the coefficient matrix) by a constant will not affect the true solution for the equation set. Sometimes this will result in diagonal dominance, however. A column of a coefficient matrix can also be multiplied by a constant. This is equivalent to the scaling of the corresponding solution value by the inverse of the constant. Thus, multiplying the kth column by c will result in the solution for the kth x-value actually to be x_k/c. These and other reordering aspects are explained in a number of references. (See, e.g., Jennings, 1977; Wait, 1979; Gustavson, 1981; Varah, 1984; Stadtherr and Wood, 1984a.) A word of caution: *partial pivoting*, the situation where pivoting is used at some intermediate stage of the elimination, can have either a positive or a negative effect on the accuracy of the solution. As has been shown (Jennings, 1977), the use of partial pivoting on symmetrical or diagonally dominant coefficient matrices can have a negative effect on accuracy.

The simple processes needed for pivoting are often shown in a *permutation matrix* , usually denoted as P and made up of zeros and ones such that there is only one nonzero entry in each column and row. Multiplying by a permutation matrix will not alter the solution to the linear system. Hence, if it is desired to interchange the second and the fifth equations in a 5 x 5 system, premultiplying by the following permutation matrix will accomplish that.

$$P = \begin{bmatrix} 1 & 0 & 0 & 0 & 0 \\ 0 & 0 & 0 & 0 & 1 \\ 0 & 0 & 1 & 0 & 0 \\ 0 & 0 & 0 & 1 & 0 \\ 0 & 1 & 0 & 0 & 0 \end{bmatrix}$$

A constant (c) times the second equation can be added to the fourth equation by premultiplying by the following matrix:

$$\begin{bmatrix} 1 & 0 & 0 & 0 & 0 \\ 0 & 1 & 0 & 0 & 0 \\ 0 & 0 & 1 & 0 & 0 \\ 0 & c & 0 & 1 & 0 \\ 0 & 0 & 0 & 0 & 1 \end{bmatrix}$$

In this simple example, the matrix shown is a lower-triangular form as was discussed earlier. The *identity matrix,* where the diagonal is filled with values of unity and other locations are null, is a special case of a permutation matrix. Any square matrix will be unchanged by pre- or postmultiplication with the identity matrix of the same order.

One final set of processes that frequently arise in the solution of linear systems of equations is noted in this brief introduction. That is the concept of *triangular factorization,* which can be shown to be equivalent to Gaussian elimination processes. It can be shown (Wait, 1979) that for any nonsingular matrix **A,** there is always a permutation matrix **P** that can operate on the original matrix to yield a form that consists of a lower-triangular, diagonal, upper-triangular form. That is,

$$PA = L^1 D U^1 \tag{4.38}$$

Here L^1 is a lower-triangular form with a unit diagonal, U^1 is an upper-triangular form with a unit diagonal, and **D** is a diagonal matrix. The diagonal matrix can be imbedded into the triangular forms by appropriate use of scaling (Wait, 1979). This gives rise to the more commonly represented form given below (Jennings, 1977; Varah, 1984):

$$A = L^1 U \tag{4.39}$$

Row and column pivoting can be applied to this system also and still yield equivalence to triangular forms. Also, if **A** is banded and does not require pivoting for the solution, the same type of triangular factorization will occur. If, in addition, pivoting is not required, L^1 and U will have the same banded form as A.

Solution of a system of linear algebraic equations as noted in Equation 4.33 proceeds directly using a triangular decomposition (Jennings, 1977). First the equation can be written as

$$L^1 U x = b$$

Then a solution is generated for a $\{y\}$ vector as follows:

$$L^1 y = b$$

This step is accomplished easily because of the triangular nature of the coefficient matrix. The solution then follows directly for x because of the triangular nature of U.

$$U x = y$$

Keep in mind that triangular decomposition is equivalent to Gaussian elimination. The decomposition approach does offer some conceptual benefits, however (Jennings, 1977).

4.2.4 Sparse Systems of Linear Equations

To introduce the concept of *sparse systems,* consider a very simple application shown in Figure 4.8. A series combination of n heaters is shown there. Assume that a gas with a constant specific heat is flowing through this bank of heaters. Thus, the temperature entering the system, gas specific heat, gas mass flow rate, and amount of heat transferred in each heater are given. The problem of finding in a sequential manner the temperatures exiting each heater is trivially easy for the situation shown. However, consider a formal simultaneous solution, as this will be instructive about some aspects of what are termed sparse systems.

Using the notation that the temperature entering the i th heater is T_{i-1} and that Q_i (kW) is transferred to the flow there, the energy balance equations are as follows:

$$m c \, (T_0 - T_1) = Q_1$$
$$m c \, (T_1 - T_2) = Q_2$$
$$m c \, (T_2 - T_3) = Q_3$$
$$m c \, (T_3 - T_4) = Q_4$$
$$\cdot \cdot \cdot \cdot$$
$$m \, c \, (T_{n-1} - T_n) = Q_n$$

This set of equations can be classified into the more inclusive *tridiagonal* matrix systems shown below. Zeros are indicated with blank entries here.

$$\begin{bmatrix} a_{11} & a_{12} & & & & \\ & a_{21} & a_{22} & a_{23} & & \\ & & a_{32} & a_{33} & a_{34} & \\ & & & a_{43} & a_{44} & a_{45} \\ & & \cdots & \cdots & & \\ & & & & a_{nn-1} & a_{nn} \end{bmatrix} \begin{bmatrix} T_1 \\ T_2 \\ T_3 \\ T_4 \\ \cdots \\ T_n \end{bmatrix} = \begin{bmatrix} b_1 \\ b_2 \\ b_3 \\ b_4 \\ \cdots \\ b_n \end{bmatrix} \qquad (4.40)$$

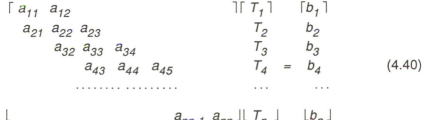

Unique heat additions

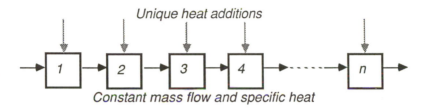

Constant mass flow and specific heat

Figure 4.8 A system of simple heaters and a mass flow of a gas through them.

If n happened to be 100, a "dense" or "packed" matrix (one in which all or nearly all coefficients have a nonzero value) would contain 10,100 coefficients. In the special tridiagonal form shown in Equation 4.40, however, there would be only 398 nonzero coefficients. Operating only with the nonzero coefficients (plus some indicators to show where the nonzero coefficients are located within the computer memory) should significantly lower the storage requirements and the computation time for a problem of this type. This statement forms the motivation for the study of solution techniques for "sparse systems" of equations. Some quick mental calculations will show that the type of system just described exhibits a greater degree of sparseness as the number of equations increases. This point is illustrated further below.

As described by Gustavson (1981), the philosophy of sparse matrix theory consists of two very important aspects: store only nonzero data and perform only operations with nonzero operands. Thus, for example, in a Gaussian elimination procedure using an appropriate sparse matrix technique, there is no need to check for zeros.

In solving a linear equation system with order 1000 (1000 equations) that has only 2000 nonzero elements, consider the savings in time and storage between dense matrix approaches (where every element is stored and processed) and sparse matrix approaches. Using conventional (dense) techniques, storage of about 10^6 words is required with almost all of the (10^6 less 10^3) zeros. The application of Gaussian elimination will require about (10^9)/3 addition and multiplication operations to solve the system. If each operation is carried out in 10^{-6} s, the dense system solution would require about 10^3 s. Use of sparse system solution techniques, however, would require less than 10^4 words to be stored and about 2×10^4 operations would be required. The computation time required for this sparse system solution would be about 20 x 10^{-6} s (Gustavson, 1981). For large systems, then, it is clear that sparse system solution techniques will save on computer storage and CPU time by several orders of magnitude compared to dense system methods.

There is no clear definition of the dividing line between what is a dense system that has some zeros and what is a sparse system. The fundamental basis is that if significant savings of computer memory and computation time can be accomplished by special handling of the zero elements, this comes under the "sparse" notation. Wait (1979) has given a suggestion on this point: if less than 10% of the elements of a system is nonzero, the system can be considered to be sparse. However, if systems with the same structure are handled numerous times, the determining fraction may be as high as 50%. It should be obvious, however, that a zero here and there does not make a sparse system.

Sparse matrix theory is too involved to be dealt with in this brief introduction. Conceptually, however, the techniques that are developing focus on the two aspects: reordering of the equations to appropriate forms and special computational techniques for the reordered equations (Jennings, 1977; Stadtherr and Wood, 1984a,b). In solving large systems of equations that could be sparse, it may be helpful to consult some of the excellent publications available on this topic for additional background (Jennings, 1977; Wait, 1979; Duff, 1980; George and Liu, 1981; Gustavson, 1981; Stadtherr and Wood, 1984a,b; Varah, 1984).

REFERENCES

Daniel, C., and F. Wood, 1971, FITTING EQUATIONS TO DATA, Wiley-Interscience, New York.

Duff, I., 1980, SPARSE MATRICES AND THEIR USES, Prentice-Hall, Englewood Cliffs, NJ.

Forsythe, G., and C. Moler, 1967, COMPUTER SOLUTION OF LINEAR ALGEBRAIC SYSTEMS, Prentice-Hall, Englewood Cliffs, NJ.

George, A., and J. Liu, 1981, COMPUTER SOLUTION OF LARGE SPARSE POSITIVE DEFINITE SYSTEMS, Prentice-Hall, Englewood Cliffs, NJ.

Gustavson, F., 1981, "Some Aspects of Computation with Sparse Matrices" in FOUNDATIONS OF COMPUTER-AIDED CHEMICAL PROCESS DESIGN, VOLUME I (R. Mah and W. Seider, Eds.), Engineering Foundation, New York, pp. 77-143.

Hang, R., 1976, "Computer Drawn Curves Using Spline Techniques," ASEE Paper, ASEE Annual Convention, June.

Jennings, A., 1977, MATRIX COMPUTATION FOR ENGINEERS AND SCIENTISTS, Wiley, New York.

Johnson, R., 1982, NUMERICAL METHODS, A SOFTWARE APPROACH, Wiley, New York.

Kolb, W., 1982, CURVE FITTING FOR PROGRAMMABLE CALCULATORS, IMTEC, Bowie, MD.

Maron, M. J., 1982, NUMERICAL ANALYSIS: A PRACTICAL APPROACH, Macmillan, New York.

Norris, A., 1981, COMPUTATIONAL CHEMISTRY, AN INTRODUCTION TO NUMERICAL METHODS, Wiley, New York.

Ralston, A., 1965, A FIRST COURSE IN NUMERICAL ANALYSIS, McGraw-Hill, New York.

Sargent, R., 1981, "A Review of Methods for Solving Nonlinear Algebraic Equations," in FOUNDATIONS OF COMPUTER-AIDED CHEMICAL PROCESS DESIGN, VOLUME I (R. Mah and W. Seider, Eds.), Engineering Foundation, New York, pp. 27-76.

Stadtherr, M., and E. Wood, 1984a, "Sparse Matrix Methods for Equation-Based Chemical Process Flowsheeting--I. Reordering Phase," Computers and Chemical Engineering, 8, pp. 9-18.

Stadtherr, M., and E. Wood, 1984b, "Sparse Matrix Methods of Equation-Based Chemical Process Flowsheeting--II. Numerical Phase," Computers and Chemical Engineering, 8, pp. 19-33.

Stewart, G., 1973, INTRODUCTION TO MATRIX COMPUTATIONS, Academic, New York.

Varah, J., 1984, "Computational Methods in Linear Algebra," Journal of Computational Physics, 54, pp. 87-94.

Viswanathan, K., 1984, "Curve-Fitting Monotonic Functions," AIChE Journal, 30, July, pp. 657-660.

Wait, R., 1979, THE NUMERICAL SOLUTION OF ALGEBRAIC EQUATIONS, Wiley-Interscience, New York.

PROBLEMS

4.1 Find the least-squares straight-line fit $y = a + bx$ for the following data. Is the straight line the best choice of an equation for this data? Show why or why not.

x	y
1.1	3.0
1.2	2.9
1.3	3.6
1.9	4.2
2.3	5.1
2.7	5.0
3.1	5.9
3.3	7.0
4.1	7.8
4.4	8.0
4.6	8.0
4.6	9.1
4.7	7.9
5.0	8.6

4.2 Consider the heating of a highly conductive object in an oven. Initially the temperature of the object is 80°F. Throughout the test the oven is 400°F. Determine an equation for the variation of temperature of the object for each of the following situations.

(a) A measurement at 2 min shows the temperature of the object is 357°F.

(b) In a separate experiment on another object, the following readings were found: temperature at 1 min = 283°F; temperature at 3 min = 384°F; and temperature at 5 min = 398°F.

(c) One final experiment is performed with a totally different set of instrumentation on another object. The temperature variation was recorded as follows: 275°F at 1 min; 372°F at 2 min; 379°F at 3 min; and 395°F at 5 min.

4.3 Develop an expression for the saturation pressure-temperature line (vapor-pressure curve) for pure steam between the temperatures of 0°C and the critical point.

4.4 Curve fit values of entropy for superheated steam in the pressure range from 1 to 3 bars and for temperatures from saturation to 500°F. Estimate the maximum error in this prediction.

4.5 It is well established that the efficiency of a solar collector varies with the amount of energy incident upon the collector. The performance of one such device is given in the table below. Shown there is the percentage of design point solar flux incident upon the collector versus efficiency of the collector, with the temperature of the collector assumed to be held constant. (The efficiency is defined as the amount of usable heat collected divided the amount of solar energy incident upon the collector.) **(a)** Find an equation giving the percentage of design flux versus collector efficiency. **(b)** Use your curve fit to estimate the percentage of design flux where the efficiency goes to zero.

% of Design flux	40	60	80
Efficiency, %	75	83.3	90

4.6 A function of two variables is as shown in the table below. Using a polynomial function, fit this data.

Values of z are underlined in the table.

	$y=$	1.0	2.0	3.0
$x=$	4.0	5.3	3.2	2.0
	5.0	6.7	9.3	8.9
	6.0	10.1	14.6	21.5

4.7 Show in detail the steps leading to Equations 4.25.

4.8 Thermophysical properties of <u>air</u> are needed in equation form for a computer routine. Using values from a heat transfer text or another standard source, determine a curve fit for the following properties at <u>standard atmospheric pressure</u> in the temperature range 200 to 2000 K: **(a)** specific heat at constant pressure; **(b)** thermal conductivity; **(c)** dynamic viscosity; and **(d)** density. After finding a correlation equation, plot the values predicted along with your original data. Calculate the maximum error of your equation. Which of these properties is most sensitive to pressure variation? Can you estimate how that property might be affected by the absolute pressure? If so, add a pressure corrector to your temperature curve fit for that property.

4.9 Solve the following system of equations with a numerical method that you compose. Do not use existing software.

$$3x_1 + 4x_2 + 2x_3 = 17$$
$$2x_1 + x_2 + 6x_3 = 22$$
$$5x_1 + 3x_2 + x_3 = 14$$

(a) Work through this with simple Gaussian elimination. **(b)** Solve by decomposing the problem with triangular factorization. In both cases, the use of the computer is encouraged, but at least the first couple of steps in the computation should be written out in detail to show the technique used.

4.10 Consider the function $\sin x - x/2 = 0$. Using a numerical technique, find any positive roots of this equation. Is the successful solution dependent upon the initial trial value of x? Compare your numerical answer to one found analytically. Answer. 1.8955

4.11 Using any software available to you or some of your own design, solve the following system.

$$Ax = b$$

where

$$A = \begin{bmatrix} 1 & 1.5 & 5.5 & 2.5 & 9 \\ 1 & 1 & 4.5 & 1 & 5 \\ 1 & 1 & 3 & 1.5 & 3.5 \\ 1 & 2 & 2.5 & 4 & 5 \\ 1 & 1 & 3.5 & 1 & 3.5 \end{bmatrix} \quad b = \begin{bmatrix} -32 \\ -20 \\ -8.5 \\ -6 \\ -12 \end{bmatrix}$$

4.12 Given an $n \times n$ system of linear algebraic equations, it is found that all equations are linearly independent except one. That is, all coefficients of one equation are found to differ from those of another equation only by a

97

multiplicative factor. If you were to proceed with Gaussian elimination without noting the relationship between the two equations, what would you expect the technique would yield for solutions? What is the theoretical solution for this situation, without resorting to an elimination scheme?

4.13 **(a)** Determine numerical values for the coefficients in the $n \times n$ matrix shown in Equation 4.40 for the system shown in Figure 4.8. Assume that each of the original equations has been divided through by the mass-flow-rate-specific-heat product, so that these factors are combined with the heat transfer terms. Are the terms on the main diagonal the dominant ones? **(b)** Solve for the temperatures for this system if $n = 35$. Take $Q_i/mc_p = i$ (°C). For example, for the ninth heater, $Q_g/mc_p = 9$°C. Also take the incoming temperature to the system to be 0°C. Solve this "duo-diagonal" system by first adapting the Gaussian elimination technique for the special tridiagonal matrix form shown in Equation 4.40 (Wait, 1979) and then apply the solution to the actual "duo-diagonal" form of Figure 4.8.

4.14 Show how you might use Newton's method to determine the value for the cube root of a number. Demonstrate this by finding $7^{0.333...}$.

4.15 Using a Newton technique, find any roots to the equation

$$y = 0 = x^3 - 7x - 6$$

in the range $|x| < 10$.   Answer. -1, -2, ...

4.16 Find a solution to the following system of equations using any numerical method.

$$x_1 + x_2 - 6x_3 = -12$$
$$x_1 - 4x_2 - x_3 = -5$$
$$3x_1 + x_2 + x_3 = 4$$

4.17 Find a solution to the following system of equations using any numerical method.

$$6x_1 + 2x_2 + 2x_3 = 8$$
$$x_1 - 4x_2 - x_3 = -5$$
$$3x_1 + x_2 + x_3 = 4$$

CHAPTER 5

ECONOMIC EVALUATION

5.1 EVALUATING THE ECONOMICS OF A PROJECT

5.1.1 Some Basic Ideas

What is the "value" of an engineering project? There can be many answers to this question. The end result of the project may be strictly for research purposes. "Does the concept work?" Some projects are designed for scale-up studies. "How does the system operate when it is made to process 5 times more product than it did in its earlier version?"

Most often, however, the guiding criterion for determining the value of a particular project is usually the economics of it. Whatever the output of the plant, there is normally a desire to produce that product at the lowest unit cost. Hence, engineering design has a large component of analysis where the costs of several alternatives are considered. This topic has received a great deal of attention in the technical press. It has been addressed in textbooks (e.g., Grant et al., 1982), government publications (e.g., Ruegg et al., 1978), and other works (e.g., EPRI, 1982).

OK, you say, I have the various costs, including *capital expenditures*, *operation and maintenance (O+M)* costs (usually including energy costs), estimates of the *replacement equipment costs* over the life of the plant, and forecast *salvage value* at the end of the planned lifetime of the whole project. The first of these is a one-time cost expended at the beginning of the project and covers the construction of the system, including the purchase of all of the equipment. The second, "O+M," is a periodic investment, often quoted on a yearly basis but sometimes given in more frequent intervals. This can include labor, expendable supplies like lubricants, filters, and packings, and energy costs. "Replacement equipment" includes expenditures for major capital equipment that must be purchased when parts wear out. It is possible to include this under O+M. Of course, "salvage" refers to money you receive when you sell the used equipment for scrap or whatever. What should be done with all of these numbers to arrive at a "bottom line" that can be used to make decisions?

A possible graphical representation of these costs is shown in Figure 5.1. Note that this is a simplification to what the real situation may be.

Differences between the situation shown and an actual project accounting are due primarily to inflation/deflation effects. These could affect both the income and the expenditures, possibly through quite different rates. Since there are historical bases to note that energy and its derivatives can inflate/deflate at different rates than other commodities and labor (there is more information given on this in Chapter 6), then it can be of value to separate the O+M costs into various components. *Taxes* also can alter the situation. Not shown is *depreciation*, a tax accountant's way of estimating the decrease in the value of property due to wear and obsolescence, which can affect the overall economic picture through the assessed amount of taxes.

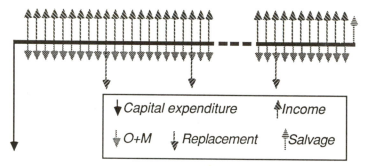

Figure 5.1 A simplified graphical representation of income and costs of a project. The various periods depicted could be years, months, or any convenient interval.

Perhaps you have analyzed a number of alternatives and the data seem like so many apples and oranges. For example, consider Table 5.1. All of the costs listed there are assumed to be for a system that has the same output of product. Which is the best buy? Ways of analyzing these types of options to determine the most economical one are described below.

Table 5.1
Cost Comparisons of Three Projects[a]

Cost Type	Alternative A	Alternative B	Alternative C
Capital	$ 1,200,000	$ 2,200,000	$ 1,900,000
O+M (and fuel)	430,000	250,000	370,000
Replacement	26,000	11,000	12,000
Salvage	9,000	12,000	11,000

a. Similar product output assumed for all. Timing of payments is as shown in Figure 5.1.

One important concept needs to be introduced before embarking on a discussion of economic analysis. That is the idea of *interest rate* and its effect on an investment. Interest, *i*, is defined as the fraction of the current principal paid per compounding period. If you put $P into a bank account at an interest rate of i%, where it may be compounded on an annual or semiannual basis, then you naturally expect the amount of money you have there to "grow." The amount of the increase depends upon the values of P and i as well as on the number of years you keep the money in the bank (n) and whether the bank pays "*simple interest*" or "*compound interest.*" Although simple interest is found infrequently in practice, Figure 5.2 shows an example of this situation, whereas Figure 5.3 shows a representation of compound interest. The accumulated amount can be calculated from Equation 5.1 for the "simple" case and Equation 5.2 for the "compound" case.

$$F_s = (1 + in) P \tag{5.1}$$

$$F_c = (1 + i)^n P \tag{5.2}$$

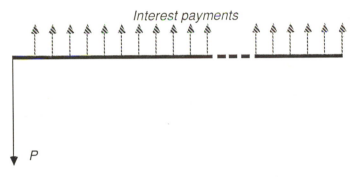

Figure 5.2 Illustration of an investment and <u>simple</u> interest situation.

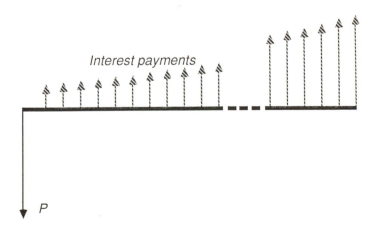

Figure 5.3 Illustration of an investment and <u>compound</u> interest situation.

Since only the compound case will be considered in what follows, the s and c subscripts will be dropped. Realize that compounding will always be done here according to Equation 5.2.

A number of variations to these basic ideas come up regularly in the economic analysis of projects. For example, the concept of *present worth* is one that uses an inverse form of Equation 5.2 as shown in Equation 5.3.

$$P = F / (1 + i)^n \qquad (5.3)$$

As was the case in Equations 5.1 and 5.2, here P is the present sum of money, and F is a future sum of money, equivalent to P at the end of n periods of time at an interest of i. This type of calculation shown in Equation 5.3 can be used, for example, to find the present value of the salvage from a project just being designed.

The ratios P/F and F/P, as well as additional ones given below, are used so frequently that they are tabulated in a number of sources like Grant et al. (1982). Routines for numerical evaluation of the ratios are available for programmable calculators.

Other discounting relationships than those already presented are also important. These include the formula for the *uniform present worth*. This is given by Equation 5.4, and it represents the amount of money at the beginning of a project that is equivalent to regular savings at regular compounding periods within the project life.

$$P = A \ \frac{(1 + i)^n - 1}{i (1 + i)^n} \qquad (5.4)$$

Here A is an end-of-period payment in a uniform series of payments over n periods at i interest rate.

Example

Consider a savings of $5000 per year for 9 years. If the annual discount rate is 8% compounded semi-annually, find the total value of the savings at the beginning of the time period.

Since the interest is compounded semi-annually, the analysis uses 18 periods at 4% each.

$$P = A \ \frac{(1 + i)^n - 1}{i (1 + i)^n} = \$5000 \ \frac{(1 + 0.04)^{18} - 1}{0.04 (1 + 0.04)^{18}} = \$63,296$$

Alternatively, Grant et al. (1982) tabulate $P/A = 12.659$ for this example.

The *uniform capital recovery* relationship is simply Equation 5.4 inverted, and this gives the regular saving amounts that are necessary to achieve a certain equivalent sum at the beginning of the project period.

$$A = P \ \frac{i \, (1 + i)^n}{(1 + i)^n - 1} \qquad (5.5)$$

5.1.2 Some Definitions

In this final section of the introduction, we give some definitions patterned after Marshall and Ruegg (1980) of some terms already introduced as well as the definitions of some terms that will be important later. Refer to this section as necessary to firm up your understanding of some of the basic terms used in the economic analysis of projects.

Benefit-cost ratio--Economic benefits expressed as a proportion of costs, where both are discounted to a present or annual value; must be greater than 1 for an investment to be economically justified on the basis of dollar-measurable benefits and costs.

Constant dollars--Values expressed in terms of the purchasing power of the dollar in a given year, usually the base year; i.e., constant dollars do not reflect price inflation/deflation.

Current dollars--Values expressed in terms of the actual prices of each year; i.e., current dollars reflect inflation/deflation.

Depreciation--The decrease in value of something when considered at two different times.

Differential price escalation rate--The expected difference between a general rate of inflation and the rate of increase assumed for a given cost component.

Discount rate--The rate of interest reflecting the time value of money, used in discounting formulas and to compute discounting factors for converting benefits and costs occurring at different times to a common time.

Discounting--A technique for converting cash flows that occur over time to equivalent amounts at a common point in time.

First cost--The sum of the planning, design, and construction costs necessary to provide a finished project or project component ready for use, sometimes called the *initial investment cost.*

Inflation (deflation)--A rise (fall) in the general price level, or put another way, a decline (increase) in the general purchasing power of the dollar.

Internal rate of return--The compound rate of interest that, when used to discount the life cycle costs and benefits of a project, will cause the two to be equal.

Major replacement cost--Any significant future component replacement, included in the capital budget, that must be incurred during the study period in order to maintain the investment at a functional level.

Nominal discount rate--The rate of interest reflecting the time value of money stemming both from inflation and the real earning power of money over time, used in discount formulas or to select discount factors for converting current dollar benefits and costs to a common time.

Operating and maintainence (O+M) costs-- The expenses incurred during the normal operation of a project as well as the costs required for corrective and preventive maintenance. Car include costs of labor, materials, and energy.

Present value (worth)--The value of a benefit or cost at the present time (i.e., in the base year), found by discounting cash flows occurring in the future to the present.

Salvage value--The net sum to be realized from disposal or sale of an asset at the end of its economic life, at the end of the study period, or when it is no longer to be used.

Simple payback period--A measure of the length of time required for the cumulative benefits net of cumulative future costs, from an investment to pay back the initial investment cost, without taking into account the time value of money.

Sunk cost--A cost that has already been incurred and should not be considered in making a current investment decision.

Total life cycle costing--A technique of life cycle costing that finds the sum of the costs of an initial investment (less salvage value), replacements, operations including energy use, and maintenance and repair over the life cycle of an investment, expressed in present or annual value terms.

5.2 SIMPLIFIED OR PARTIAL METHODS OF ECONOMIC EVALUATION

5.2.1 Introduction

A number of commonly used methods of economic evaluation do not take into account all of the important variables that can enter into the analysis. Two of these methods are mentioned here. The reason for including them is not to advocate their application on a regular basis. Rather it is for a two-fold purpose: to introduce the terminology so that you can recognize the concepts when someone else refers to them and to list some limitations of the techniques. There may be times when one of these methods can be used to give comparative ranking of alternatives. Since more comprehensive methods are easily handled, this is not recommended, however.

5.2.2 Return on Investment (ROI) Method

The ROI method uses the average annual <u>net</u> benefits (income less expenditures) and ratios this value to the original book value of the project. This is shown in Equation 5.6. This approach is similar to one used traditionally in mining enterprises termed Hoskold's method (Grant et al., 1982).

$$ROI = \frac{Average\ annual\ net\ benefits}{Original\ book\ value} \times 100 \qquad (5.6)$$

Advantages associated with the ROI method include ease of use. The business world uses the concept a great deal. *Disadvantages* include the fact that the timing of cash flows is not included, and this can make a great deal of difference in the comparison of projects. For example, if two investments are considered with one paying large benefits for four years and another paying slightly smaller benefits over seven years, ROI might favor inappropriately the first investment possibility. Another disadvantage, particularly when applied to complicated process plant projects that are frequently addressed by mechanical and chemical engineers, is that ROI calculations require the "original book value" of a design. This is an ill-defined term and is handled in different ways by different organizations. Even though this technique is often incorrectly applied, many accountants still use it frequently.

Example

Consider the following data and calculate the associated ROI:

Original book value = $90,000
Expected life= 15 years
Annual depreciation (straight-line method) =
$90,000 / 15 = $6000
Yearly O+M = $1500
Yearly product value (income) = $30,000.

This situation is plotted in Figure 5.4. Using Equation 5.6, the ROI is found, and its value is given below.

$$ROI = \frac{\$30,000 - (\$6000 + \$1500)}{\$90,000} \times 100 = 25\%$$

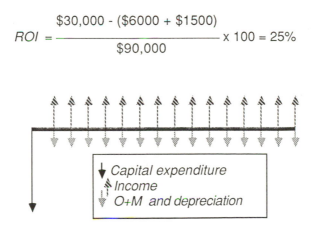

Figure 5.4 Nondiscounted cash flow diagram for the ROI example.

5.2.3 Payback Method

Where the ROI method seeks an annual value or "interest" on investment, the payback method is used to find a simply defined number of years to "pay back" the investment from the receipts of the project. In a sense, it finds the number of years required to "break even" on a project. Usually the calculation is performed on a before-tax basis and without including the "time value of money" (interest). The payback period (PP) is often calculated as shown in Equation 5.7, and it should be noted that there is a close relation between the resulting value and the inverse of the value found from Equation 5.6. While the payback method is used by a large number of industrial organizations, it is almost always applied in conjunction with other approaches.

$$PP = \frac{First\ cost}{Average\ yearly\ benefits - Average\ yearly\ costs} \tag{5.7}$$

There are a number of *advantages* with the payback as a means of economic evaluation. This technique is usually a good one to use when very short term investments are being considered. Payback is often used for assessing the likelihood of a favorable investment from a project with an uncertain life.

Disadvantages of the technique are shared with the ROI technique. Since the method does not give consideration to cash flows beyond the payback period, the efficiency of an investment over the entire project life is not assessed. (Of course, a payback comparison might be made with good accuracy of two projects with the same life and year-to-year performance that does not vary.) Since interest rates and the time value of money are not included, there can be erroneous results from an economic analysis, as shown in the example below. The discount rate can be included in the payback analysis, but often this variable is not.

Example

Consider two investment possibilities that have first costs of $220,000 each. The first project, (**a**), has yearly net benefits for the first and second years of $70,000 and $150,000, respectively. Yearly net benefits for the first and second years, respectively, for the second investment possibility, (**b**), are $140,000 and $80,000. This is shown in Figure 5.5. Determine the payback, without including interest effects, and comment on whether or not this means of evaluation is a good one to use in this situation.

Using the first two years' data with Equation 5.7, find that both have the same payback period--two years. That is,

$$PP = \$220,000 / (\$220,000 / 2) = 2 \text{ years.}$$

Hence, simple payback analysis would indicate that these investment opportunities are equivalent. If, however, a reasonable discount rate is included, say 10% compounded annually, and the total project benefit at the end of the payback period just calculated is analyzed, the conclusion would be different. Total income from option (**a**) at the end of two years would actually be $249,700, while the corresponding amount for option (**b**) would be $257,400. The discount rate can be included in a payback analysis, but often this is not done.

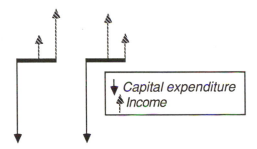

Capital expenditure
Income

Figure 5.5 The cash-flow diagrams for two situations used in analyzing payback.

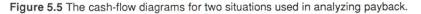

5.3 LIFE CYCLE COSTING

More comprehensive techniques for economic analysis generally do not suffer the shortcomings noted in the two shortcut methods outlined above. There are a number of these more comprehensive ways of evaluation. Included are the Savings to Investment Ratio, Internal Rate of Return, Discounted Payback, Net Benefits, Life Cycle Costing (Ruegg and Marshall, 1981) and Equivalent Uniform Annual Cost (Grant et al., 1982) analyses. Both the LCC and Net Benefits methods are often used to extract similar types of comparisons. Applications include situations when the end products can be directly compared. This is the situation that occurs in evaluating the design options for a process plant. The Savings to Investment Ratio and Internal Rate of Return methods give information that is of value in comparing projects that have quite different end results (perhaps like the plant construction situation faced at the corporate level of a widely diversified company in choosing between a new widget factory or a new power plant), and, in this way, they are similar to one another and quite different from the LCC and Net Benefits methods.

Focus here will be on only one of the more comprehensive methods: Life Cycle Costing (LCC). This does not imply that this is the only method that can be used. In fact, it is in this variety of approaches that many strong biases for or against a particular method are found among engineering economists. Instead of claiming one technique is preferred over another, the approach taken here is to introduce one representative technique that, when properly used, can give a good economic picture of trade-offs between projects whose output is the same. As noted above, LCC will not give information on the rate of return of an investment. This information could be of considerable benefit when comparing projects whose outputs are diverse or for evaluating the investment value of a single project.

Consider two alternatives for plant modifications to accomplish the same end result--perhaps the manufacture of a certain amount of a commodity. Assume that these options have different initial costs, different O+M, different utilizable lifetimes, and different energy requirements. In short, this could be the "apples and oranges" situation mentioned earlier in this chapter, with the one restriction that the product output is the same over the same time period. The LCC approach allows a rational economic basis to compare the two alternatives. Equation 5.8 gives the basis for the calculation of the LCC.

$$PW(Total\ cost,\ LP)\ =\ PW(Initial\ cost,\ LP)\ +$$

$$PW(O+M,\ LP)\ +\ PW(R,\ LP)\ -\ PW(S,\ LP) \qquad (5.8)$$

where $PW(X,LP)$ denotes the present worth of item X over the life of the project, LP; $O+M$ is the discounted operation and maintenance cost not including energy costs; R is the discounted replacement cost for capital equipment; E represents a discounted energy cost; and S is the discounted salvage value at the end of the project. All of these costs must be estimated at the same point in time, so appropriate discounting must be taken into account. Energy costs are separated from other $O+M$ costs to be able to include effects of different values of inflation/deflation on these two quantities, if desired. For a simplified analysis, the inflation/ deflation effects could be neglected and $O+M$ and energy could be combined. This will be the approach used in the example described below.

If the product output of the two systems is the same, then the product income does not have to be brought into an LCC analysis. (However, if a corporate planner must decide to put funds into this project or into one in quite a different plant somewhere else, this may not be the case. He/she may be concerned about the output value of the plants compared to their respective investment.) For simplicity, the effect of taxes will not be included here. In the recent past, the tax laws have been changing often.

The first step in the LCC analysis approach used here is the definition of a project life cycle. As hypothesized above, the two project options have different estimated lifetimes. Somehow these have to be brought to a common basis. Usually this is done by assuming that the shorter life system is replaced at appropriate times for realistic costs, and the end of the analysis period will find that option with a significant service life remaining. This is shown in Figure 5.6. As a result, the replaced system will probably have a significant salvage value at the end of the analysis period because this system will not be at the end of its life.

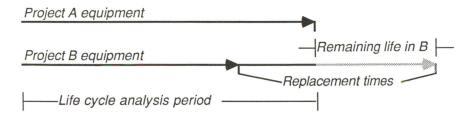

Figure 5.6 A possible way of comparing two projects with equipment that has different lifetimes is to set the life cycle analysis period equal to the life of the longer lived project (A, above). Assume that replacement of the shorter lived equipment (B, above) takes place as many times as necessary to make the combination have at least as long a life as A. Of course, a higher than usual salvage value should then be assessed for B at the end of the analysis period.

Note one extremely important point here. This analysis has an initial step that involves estimating the project life and using this value in the calculations that follow. The project life may, in fact, be nearly impossible to assess at the onset. For this reason, many engineering economists favor approaches that treat the project life as an unknown parameter. Total project costs can then be studied as a function of this variable. These approaches can be easily applied when project design alternatives are between ones with similar product outputs.

Table 5.2
Comparison of Two Projects' Costs

Item	Project A	Project B
Period of analysis	20 years	20 years
Normal life	20 years	15 years
Capital investment	$510,000	$240,000
O+M and fuel	$ 10,000[a]	$ 40,000[a]
Replacement	0	$120,000[b]
Salvage	0	$ 80,000[c]

a. Each year. b. 15th year. c. 20th year.

Consider the nondiscounted cash flow information for two systems shown in Table 5.2. Take the period of analysis to be the project life of the longer lived project (A). Hence, the shorter lived project (B) will require some major equipment to be added at the end of its useful life, assumed here to be 15 years. Note that the replacement cost is not the same as the initial capital investment because it is assumed that some of the first costs will not have to be repeated. The percentage that would have to be repeated varies from project to project and could be 100% in some situations. Note also that there is a significant salvage value for project B at the end of the analysis period because, it is assumed, there is still significant value in the B equipment left at that point. A possible way of handling this "artificial" salvage is to take a linear relation between time and replacement costs. In the case shown, there is still 2/3 of the replacement cost life left at the end of the project. "Conventional" salvage values have been neglected; but some realistic values, at most usually a few percent of the original capital investment, could be included.

Approximate cash flow diagrams for these two projects are shown in Figure 5.7. As can be seen from there as well as from Table 5.2, the two projects represent the comparison between a higher first cost, lower operating cost project (A) with a lower first cost, higher operating cost project (B). Which project is best economic investment? This is not obvious to most people without some analysis.

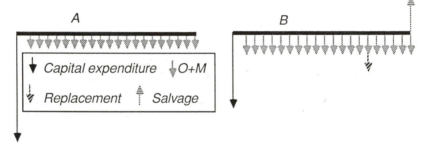

Figure 5.7 Cash flow diagrams for two projects that are to be compared for their total life cycle costs.

To do the analysis, it is necessary to make the comparisons and calculate the present value of the various cash flows. Use is made of some of the discounting relationships introduced earlier and an assumed value for the <u>discount rate of 10%</u>. The initial investment values translate directly to present values, as these expenses will be put into the project now. A present value for the salvage income as well as for the replacement expenditure can be calculated from the single present worth evaluation, described above in Equation 5.3. (While the equivalent costs at the <u>beginning or the end</u> of the project could be evaluated, the beginning of the project is the basis for calculations here.) O+M plus fuel costs must be handled differently than single expenditures or income. These expenditures are tabulated on a yearly basis, so a uniform present worth evaluation must be used to find the lump sum equivalent expense at the beginning of the project. To do this, use a form of Equation 5.4. Results of all of the calculations for the cash flows indicated in Table 5.2 are shown in Table 5.3. Inflation effects are neglected.

Table 5.3
Present Values of Project Costs in Table 5.2[a]

Item	Project A	Project B
Capital investment	$510,000(-)[b]	$240,000(-)
O+M and fuel	85,140(-)	340,560(-)
Replacement	0	28,680(-)
Salvage	0	11,920(+)
Total present value	595,140(-)	597,320(-)

a. Based upon a 10% discount rate. **b.** (-) denotes an equivalent cash outlay.

With the numbers given in Table 5.3 at hand, the decision about which of the projects is the "best buy" follows directly, although in fact this case shows almost equivalent values. Here a decision may be made on other grounds than strictly economic--perhaps considering technological factors.

Keep in mind what has been assumed and what has been found. It was assumed that each of the projects had the same amount of useful output. If the project produces power, a chemical product, or whatever, the receipts for sales will be the same. In addition, it should be noted that virtually all of the values entered in Table 5.3 are expenditures. Salvage is the only exception as it is the equivalent to money in our pocket that is realized at the end of the project. Because that "income" is developed after 20 years where the discount rate is 10%, the present value $11,920 is the equivalent to $80,000 then.

The best buy is the project with the lowest life cycle cost. In the case shown, each of the options has approximately the same cost, and, hence, each is nearly equal. Consideration should be given to other factors in this case for possibly helping with the decision as to which to use. We have not included inflationary effects in our calculations. We will return to this example in the next section to incorporate a factor for this. If this does not distinguish a clear "winner," then we would probably need to use technological insights in picking one of the options. The example shown in Tables 5.2 and 5.3 is unusual in that the options yielded similar costs. Normally, at this point, there would be more distinction between the two.

Remember that a cost/benefit ratio or a rate of return on investment has not been found. That calculation is more involved than the present one and would require knowledge about the value of the item(s) produced.

In summary, it has been shown that total life cycle costing is a convenient method to employ in making economic comparisons between project possibilities. With information about the single and reoccurring costs associated with a project, the "best buy" can be determined.

5.4 THE EFFECTS OF INFLATION

5.4.1 Introduction

Throughout history, prices of products have varied considerably. The prices of some items have been greatly affected by their scarcity (such as the depleting of a limited resource like diamonds) or the development of technology (such as the significant breakthroughs in electronics that have impacted the prices of products derived from that field). In most situations, though, the prices of objects are affected by the more global impacts of inflation or deflation. Under these influences, the prices of nearly all commodities and labor go up or down almost in unison. While this topic is given more attention in Chapter 6 when specific aspects of component cost estimating are addressed, it is sufficient to note that inflationary/deflationary (in what follows the term "inflation," only, is used, while in most situations the appropriate reference is to "inflation/ deflation") effects can have some important impacts on the prices of products. This, in turn, can have significant effects on the "bottom line" of a project. Effects of inflation included in the economic analysis of a project may completely change the perceived viability of that project. A basic method for incorporating inflation into a project analysis is discussed in what follows.

Before beginning this discussion of the analysis of inflation, some cautions should be noted. First, effects of inflation will come into the analysis as estimates of what will be the situation during the life of the project. As with all "crystal ball gazing," this can be highly inaccurate. It is often necessary to treat the price change rate as a parameter, encompassing the range that you think is most probable. This parametric evaluation, sometimes called a *sensitivity analysis*, can be extremely valuable in illustrating the importance of the unknowns.

Second, as will be noted in Chapter 6, the rates of all commodities do not change in nearly the same way. While many do, there can be exceptions. One good example, pertinent to the thrust of this text, is energy. Since the energy supply can be affected by political factors as well as the genuine availability of energy (if there is a true growing or shrinking of the energy stockpiles), energy prices may escalate or de-escalate at a different rate than other prices. The methods for handling two simultaneous rates of inflation are not covered here. Information can be found elsewhere (Turner and Case, 1979; Baum, 1984).

5.4.2 Analysis Principles

In investment analysis, an approach that is often followed to incorporate inflation effects is to assume that all costs and revenues inflate at the same rate (Ruegg, 1977). With this assumption, all prices remain constant in real terms, and this greatly simplifies the analysis.

Two factors not covered here could bring this constant price assumption and its effects into question. One of these is taxes. The real after-tax return to an industrial organization may be substantially changed by inflation, even if pretax investment costs and income are assumed to change equally with inflation. Further discussion of the implications of this are given by Ruegg (1977). A second factor that can be affected by the assumption of equal inflation rates is that the source of investment funds does not importantly influence the outcome of the analysis. This may not be the case. Both of these factors are neglected in what follows.

The key element in the analysis that is a change from what was described in Section 5.3 is that future expenditures, and income, if tallied, must have the inflation rate effects included. For example, if the estimated discount rate is denoted by d and the estimated inflation rate is given the symbol I, the expense of O+M costs must be escalated at the rate $(1+I)^n$ and discounted at the rate $(1+i)^n$. Thus, the appropriate factor, in comparison to that given in Equation 5.4, to account for periodic payments or revenues when inflation is present is given by Equation 5.9.

$$P = A \left[\frac{1+I}{1-I} \right] \left[1 - \frac{(1+I)^n}{(1+i)} \right] \qquad (5.9)$$

Example

Reconsider the project comparisons shown in Tables 5.2 and 5.3 but now include the fact that the O+M and fuel costs are assumed to be escalating at 5% per year. Determine the life cycle costs now.

Reiterate much of what was given in Table 5.3 in Table 5.4, changing only the costs for O+M and fuel, and total each of the columns with the new information. Now Project A is the clear "winner."

Table 5.4
Present Values of Project Costs in Table 5.2[a]

Item	Project A	Project B
Capital investment	$510,000(-)[b]	$240,000(-)
O+M and fuel	85,140(-)[c]	340,560(-)[c]
Replacement	0	28,680(-)
Salvage	0	11,920(+)
Total present value	595,140(-)	765,480(-)

a. Based on a 10% discount rate. b. (-) denotes a cash outlay. c. O+M and fuel assumed to be inflating at 5% per year.

REFERENCES

Baum, S., 1984, "Engineering Economy and the Two Rates of Return--Mixed Mode Computations," in INFLATION AND ITS IMPACT ON INVESTMENT DECISIONS (J. Buck and C. Park, Eds.), Industrial Engineering and Management Press, Atlanta.

EPRI, 1982, TAG™, TECHNICAL ASSESSMENT GUIDE, P-2410-SR, Electric Power Research Institute, Palo Alto.

Grant, E., W. Ireson, and R. Leavenworth, 1982, PRINCIPLES OF ENGINEERING ECONOMY, SEVENTH EDITION, Wiley, New York.

Marshall, H., and R. Ruegg, 1980, ENERGY CONSERVATION IN BUILDINGS: AN ECONOMICS GUIDEBOOK FOR INVESTMENT DECISIONS, NBS Handbook 132, National Bureau of Standards.

Ruegg, R., 1977, "Economics of Waste Heat Recovery," Chapter 3 in WASTE HEAT MANAGEMENT GUIDEBOOK (K. Kreider and M. McNeil, Eds.), NBS Handbook 121, National Bureau of Standards.

Ruegg, R., and H. Marshall, 1981, "Economics of Building Design," Solar Age, July, pp. 22-27.

Ruegg, R., J. McConnaughey, G. Sav, and K. Hockenbery, 1978, LIFE-CYCLE COSTING, A GUIDE FOR SELECTING ENERGY CONSERVATION PROJECTS FOR PUBLIC BUILDINGS, NBS Building Science Series 113, National Bureau of Standards.

Turner, W., and K. Case, 1979, "Economic Analysis of Energy Proposals," FALL IE CONFERENCE PROCEEDINGS, pp. 47-53.

PROBLEMS

Note that the discount and inflation rates given in the problems below are strictly hypothetical and do not necessarily relate to a real situation. Problems 5.1-5.10 do not include inflationary effects.

5.1 What is the present value of a replacement cost of $2500 that will occur at the end of 10 years if the discount rate is 10%?

5.2 What is the present value of a yearly routine maintenance cost of $2000 over 20 years **(a)** if the discount rate is 10%? **(b)** If the discount rate is 5%?

5.3 What is the annual value equivalent of a replacement cost of $2500 expected to occur in 10 years, **(a)** assuming a discount rate of 10%? **(b)** assuming a discount rate of 7%?

5.4 What will be the net present value savings, before taxes, of renovating the furnace in an industrial plant if the investment costs are $50,000 and the net annual fuel savings are $9000 for 15 years. Assume there is no inflation/deflation and **(a)** that the discount rate is 10%. **(b)** How much does the net present value savings change if the discount rate is 8%?

5.5 What is the maximum investment cost that could be incurred for the following energy conservation project in order to avoid a net loss? It is proposed to replace an existing electric resistance HVAC system with a heat pump. The yearly electrical cost for the resistance heating system is approximately $2000. An energy savings of approximately 50% is estimated with the heat pump. You may assume the following to be true: the existing electric resistance system has no salvage value when replaced; the remaining life of the existing system (if not replaced), the life of the heat pump, and the life of the house are all estimated to be 25 years; maintenance and repair costs for the two systems are identical; the discount rate is 10%; inflation/deflation effects can be neglected.

5.6 You are asked to evaluate the effects of insulating a gas-fired heater used on the process line. The cost of adding the insulation is estimated to be approximately $7000, and it is estimated that $800 in fuel will be saved each year. The discount rate is 10% and the expected remaining life of the heater is 10 years. Neglect any salvage value. Is it cost-effective to add the insulation? What will be the net savings or losses?

5.7 Consider the two alternatives shown in Table 5.5. Which is the "better buy" if the project system output is the same as before and the discount rate is 9%?

Table 5.5
Comparison of Two Projects' Costs

Item	Project A	Project B
Normal life	15 years	10 years
Capital investment	$ 10,000	$ 7,000
O+M and fuel	$ 3,000[a]	$ 5,000[a]
Replacement	$ 9,000[b]	$ 5,000[c]
Salvage	$ 500[d]	$ 350[d]

a. Each year. b. 15th year. c. 10th year. d. At end of normal life.

5.8 Consider the following information about some process system equipment. It is estimated that the system will require $20,000 of replacement expenditures every 10 years to keep it functioning. The O+M and fuel costs are estimated to be $100,000 per year. A forecast life remaining on the system is 25 years, and the discount rate is 10%. Neglect inflation/deflation effects. What is the present total life cycle cost of the system?

115

5.9 A totally new equipment replacement for the system noted in Problem 5.8 is being contemplated, and you are asked to evaluate the relative benefits of doing this. The new system will produce exactly the same amount of product as the old system but will cut down on the yearly operational costs. It is estimated that the decrease in the yearly costs will be 12.5% for a capital investment of $100,000. Is this a good investment?

5.10 What is the present value of a replacement cost of $100,000 that will occur at the end of 20 years if the discount rate is 8%?

Problems 5.11-5.20 are Problems 5.1-5.10 revisited, now including inflationary effects.

5.11 What is the present value of a replacement cost of $2500 (based upon today's prices) that will occur at the end of 10 years if the discount rate is 10% and the rate of inflation for this product is assumed to be 4%?

5.12 What is the present value of a yearly routine maintenance cost of $2000 over 20 years **(a)** if the discount rate is 10%? **(b)** If the discount rate is 5%? In each case consider the effects of a 1% and a 5% inflationary rate.

5.13 What is the annual value equivalent of a replacement cost of $2500 expected to occur in 10 years, **(a)** assuming a discount rate of 10%? **(b)** assuming a discount rate of 7%? In both cases assume the inflation rate is 4%.

5.14 What will be the net present value savings, before taxes, of renovating the furnace in an industrial plant if the investment costs are $50,000 and the annual fuel savings are $9000 for 15 years. Assume that the inflation rate is 10% and **(a)** that the discount rate is 10%. **(b)** How much does the net present value savings change if the discount rate is 8%?

5.15 What is the maximum investment cost that could be incurred for the following energy conservation project in order to avoid a net loss? It is proposed to replace an existing electric resistance HVAC system with a heat pump. The yearly electrical cost for the resistance heating system is approximately $2000. An energy savings of approximately 50% is estimated with the heat pump. You may assume the following to be true: the existing electric resistance system has no salvage value when replaced; the remaining life of the existing system (if not replaced), the life of the heat pump, and the life of the house are all estimated to be 25 years; maintenance and repair costs for the two systems are identical; the discount rate is 10%; inflation rate is 8%.

5.16 You are asked to evaluate the effects of insulating a gas-fired heater used on the process line. The cost of adding the insulation is estimated to be approximately $7000, and it is estimated that $800 in fuel will be saved each year. The discount rate is 10% and the expected remaining life of the heater is 10 years. Neglect any salvage value. Is it cost-effective to add the insulation? What will be the net savings or losses? Fuel costs can be assumed to be inflating at 6% per year.

5.17 Consider the following information about some process system equipment. It is estimated that the system will require $20,000 of replacement expenditures every 10 years to keep it functioning. The O+M and fuel costs are estimated to be $100,000 per year. A forecast life remaining on the system is 25 years, and the discount rate is 10%. Assume all costs escalate at 5% per year. What is the present total life cycle cost of the system?

5.18 A totally new equipment replacement for the system noted in Problem 5.17 is being contemplated, and you are asked to evaluate the relative benefits of doing this. The new system will produce exactly the same amount of product as the old system but will cut down on the yearly operational costs. It is estimated that the decrease in the yearly costs will be 12.5% for a capital investment of $100,000. Is this a good investment?

5.19 What is the present value of a replacement cost of $100,000 that will occur at the end of 20 years if both the discount and inflation rates are 8%?

5.20 Consider the two alternatives shown in Table 5.5. Which is the "better buy" if the project system output is the same as before, the discount rate is 9%, and the inflation rate is 5%? Should you include an inflationary factor on the replacement cost of Project B if you use a 15-year project comparison?

CHAPTER 6
PRELIMINARY COST ESTIMATION

6.1 INTRODUCTION

It is a very unusual design where the cost of the end product is not important. Even in the height of the race for the manned landing on the moon when very high value was placed on reliability, success, and expediency, estimates of the "bottom line" were still needed. With the world market becoming more competitive, the search for designs that are optimized to be both functional and economical is a fact of life. This is certainly true for the design of thermal systems.

As was discussed in the previous chapter, a number of factors make up the "cost" of a system. In general, there are capital costs, fuel and/or power costs, other operating and maintenance costs, taxes, interest, as well as a number of other factors. As discussed in Chapter 5, the prediction of each of these aspects can be quite involved. A sea of cost estimators and accountants may in fact be needed to predict all of the cost ramifications of a given design. It is this detailed information that would probably be required for company management to make a decision about constructing a new plant or making a major investment in a new process to replace an old one. Several excellent books address many of the important aspects of the full analysis. In addition to the references noted in the previous chapter see, for example, Woods (1975) and Humpreys and Katell (1981).

While these in-depth analyses are very critical in the business aspects of the engineering world, they are something that the preliminary design engineer rarely has to do. Instead he or she will usually be concerned with cost and performance trade-offs in designs to accomplish a specific function. Of course, this function could range from one with very small cost ramifications to one that has extremely large investment implications.

Even a focus on the capital investment only can be a major undertaking. For example, if the purchase of a small steam generator system is desired, the estimate of the capital cost of that alone can be quite involved. A variety of design approaches could satisfy the end need, and each of these may have its own cost/performance aspects that need to be determined. (If all of this is known at the onset, one of the accountants could probably make the decision.) Of the cost factors noted above, the ones that will be given attention here are the first two: equipment and fuel costs.

Short of having turn-key bids for the complete system with guaranteed specifications, some cost estimates will have to be made. While the costs of the individual items of major equipment are important, they are often given on an FOB[1] basis. Costs of shipping to the site and labor charges for erection will have to be added. Both of these factors can vary considerably with the location of the installation due to distance and difficulty of transporting the equipment and the cost and productivity of the local labor. There will probably be some indirect (management and other overhead) costs that need to be considered also. If the construction is to be performed by a company other than your own, outlay for profit will have to be considered. Also, do not forget the possible need for additional buildings, all the minor equipment and piping, and its cost of installation, the attendant instrumentation, and...., and.... Although the list seems endless, there is hope for some possibility of getting a direct handle on cost trade-offs.

Often the situation is such that a critical factor in the design analysis is the cost of the capital equipment items. To see why this can be so, accept the hypothesis that an emphasis on "preliminary" cost estimation will be of value. In this context the techniques used will be much simpler and less comprehensive. Do not expect this to give a total cost that you could stake your job on. It will only give one-place accuracy in the total estimate in an absolute sense. As the total cost goes up, the larger will be the amount of the dollar error.

This approach will have a number of benefits. First, an aspect you are seeking for your work is a way to incorporate important cost trends. For example, if heat exhangers are made larger in an attempt to decrease the fuel costs for a particular design, it would be nice to have a cost function that reflects the size effects of heat exchangers to use in your analysis. Also, if the cost factors for each of the major components in a design are known, then this can be valuable in considering trade-off options with the more costly items. Finally, there are shortcut techniques (approximate, of course) where the total costs for a plant or process can be estimated from the major equipment costs. Even the detailed preliminary cost estimation approaches deal with the cost of major components. All in all, looking at the major equipment costs during the preliminary design turns out to be a good compromise between performance (getting answers) and investment (spending your time).

If you are in the business of designing the same kind of equipment over and over again, it will not take you very long to realize that a bank of cost data would be very handy. Not only would this include the costs of

---

1 WEBSTER'S NEW WORLD DICTIONARY defines F.O.B. as "free on board: used in quoting prices of goods at the place of manufacture, not including transportation charges." It certainly does not include any installation.

the materials you purchase and the costs of your product line, but it should also include costs of competing products. The major producers of equipment in the world (companies like GE and GM) have numerous files of cost information. If you are just starting out in design, or if your business encompasses a large number of new concepts, you may not have data on specific components that would be of value to your work. What are your options?

If price information is needed, there is no question that the best source is the vendor of the particular product(s). Often the vendor will furnish a firm quote that will be valid for a specific length of time. Hence, if the product is purchased during that time, be assured that the cost will be as quoted.

While vendor contacts represent the height of accuracy on a specific piece of equipment, there are some drawbacks to this for someone who is doing systems' modeling for design. One problem is that contacts of this sort take a great deal of time. A second shortcoming is that what is most often of value in simulation work are generalized price estimates. For example, it may be of value to know how the cost of centrifugal pumps made of carbon steel varies as function of rated flow and pressure rise. This information can be tabulated by making many contacts with vendors. (See the Problems section.) This has to be done only one or two times to see that the process can be quite time consuming and data can be difficult to correlate. For these reasons, it is desirable to have access to tabulations of cost information.

Information on component costs have been tabulated in a variety of ways, and these are available in a number of sources. A description of one type of summary of some of this data follows in this chapter. Since summaries are historical, they have to be made current in some way to be of value. The typical ways that this is handled are also described here. Tabulations of cost information data are given in Appendix D.

It is valuable to examine ways that major equipment prices can be used for inferring the costs of the total project. An approximate method to accomplish this is given in Section 6.2.3. In Section 6.3, examples of the use of the data presented in Appendix D are given.

Finally, some estimates of fuel costs in the future are given Section 6.4. This information is critical in predicting life cycle costs of thermal systems.

6.2 THE TECHNIQUE

6.2.1 Size Effects on Cost of Equipment

Summaries of equipment and component data found in the literature are usually presented in one of three ways. Cost information from vendors for specific models of items is the most detailed. This has already been discussed.

120

A handy and comprehensive way to present data is to give a detailed plot or mathematical function of the variation of costs with important parameters. These are found in number of sources (Guthrie, 1969; Peters and Timmerhaus, 1980; Vatavuk and Neveril, all; Woods, Anderson, and Norman, all). See also the numerous listings of U.S. government reports given in Appendix E with the numbers that include "ANL/CES/TE." The most complete form of these representations is where the cost and all critical performance information is combined in a single plot. An example of this is shown in Figure 6.1. It is apparent that a great deal of information about the price of wind power generators can be determined from this plot. Usually this amount of detail is not conveniently available.

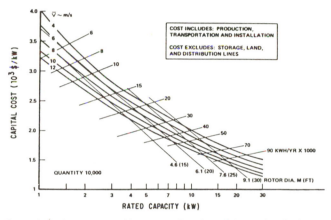

Figure 6.1 Cost of wind power machines as a function of the major design variables. Data represents prices current as of 1975. (From Yeoman, 1978.)

Cost is most often given as a gross function of a single parameter, perhaps with other characteristics fixed. Always note the applicable date that the information was gathered (more is given on this in Section 6.2.2), and also be concerned about whether or not there may have been technological or marketing developments that might have a major impact on the price functions since they were formulated. For example, the price of wind power machines might have been affected considerably by the major market stimulation that took place in the late 1970s and the early 1980s.

A third way that data is often presented in an approximate, compact format is in the so-called *exponential form*. Consider a plot of data similar to Figure 6.1 but in the much simpler relationship between size and cost as shown in Figure 6.2. If the data fall on a straight line, or a nearly straight line, on a log-log plot, then it is known that a power function in the size parameter will represent the variation. To describe this completely, the slope of the line and one point on the line is needed. As shown in the figure, *m* denotes the *slope* and the *reference size and costs* are given. If

121

these latter values are taken to be S_r and C_r, respectively, then the equation of the cost curve is given as

$$C = C_r (S / S_r)^m \qquad (6.1)$$

where C is the cost for a size of interest, S. The factor m usually falls in the range of 0.5-1.0 (values less than unity give rise to what is called "economies of scale"), but sometimes m is greater than 1.0. Thus, a complete tabulation of data for a particular piece of equipment could contain the following information:

- A complete, physical description of the equipment.
- C_r, S_r, and m.
- The range of sizes correlated: S_{min} and S_{max}.
- Correction factors for typical options for the equipment.
- Some indication of error in the correlation (how accurately the correlation describes the cost information).
- Date that data were current.

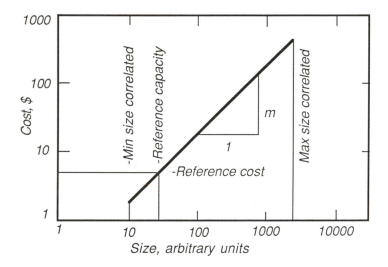

Figure 6.2. An example of the approximate cost function used here for a piece of equipment showing variables pertinent to size effects.

A large number of tabulations for pertinent pieces of equipment are given in Table 3 of Appendix D. An attempt has been made to compile the first three pieces of information listed for every item. Some data are limited, however. Correction factors for optional configurations are given in some cases. Error limits in the correlations are not listed. In some instances, these can be found by consulting the source cited in the table. All data have been updated as will be described in the next section. Some sources list the effective date of their data, others do not; so there have been base estimates made in some cases. Sometimes data that is too old may not be desirable even though it has been "updated."

When information on a specific item is not given, many times a reasonable cost function can be inferred from a related item that is tabulated. If no information is given, then take $m=0.6$.

One final point. Do not be too surprised that there may be a large error on a given correlation. Consider a hypothetical tabulation of the cost of "four-door, sedan automobiles." With this brief description, it is to be expected that all foreign and domestic vehicles will show a wide range of costs. In general, the more complete the description of the equipment and the more specific the category, the less error will be involved. All tabulations in Appendix D are very approximate.

6.2.2   Updating Historical Data

Once the cost of a piece of equipment is found, either from a detailed plot of data, a logarithmic data fit, or a specific tabulation of data, the question should then be asked: Is this cost pertinent for the desired application date? Since all data are historical and costs do change, this is a valid question. In the period 1970-1980, for example, the costs of most items changed considerably.

Considering why costs change will help to see how historical data might be updated. In general, prices of manufactured items can be influenced by four dominant factors. One of the primary factors is a valuable key to the updating of data.

By far the largest influence on the market price of an item is the effect of inflation/deflation trends. While the inflation/deflation trends are, in turn, driven by a number of stimuli, the effects tend to be widespread in the economy. This means that the definition of relevant kinds of price indices can be valuable in reflecting a record of inflation/deflation trends. More is given on this below.

A second factor that influences prices is technology. When the transistor replaced the vacuum tube, a whole range of electronic products became available at significantly lower prices. It is usually the situation that significant technological developments have an effect of lowering the price of affected products. These factors are very product specific.

The third driver on price changes is marketing competition. When other manufacturers challenged Sony in the Video Cassette Recorder (VCR) field, the prices of these items started a significant downward trend. While there were still VCRs that cost as much as the original Sony units long after the market became more competitive, the newer expensive units had a much greater performance capability than did the originals.

Finally, market stimulation can affect prices. While this can be related to competition, sometimes it is quite distinct. If only a small number of a product is made and sold each year, the unit cost will usually be higher than if mass production can be used. Government tax incentives on energy conservation products were an example of this.

Which of these four influences is important to the prices of thermal systems and components? While all four bear on this field, there is no question that the inflation/deflation factors by far carry the most weight except in isolated instances.

Now consider the most important factor that affects the time variation of the cost of equipment in thermal systems. For the sake of illustration, assume that all inflation/deflation factors occurred "across-the-board." That is, assume that when the price of one product increased, all other items went up by the same amount, including the price of automobiles, wrist watches, natural gas, and wages. Then if someone monitored the price of any or all of the marketed products, he/she could give a plot of the historical variation in prices, probably normalized in some way. Call this normalized record an *index*. If it is further assumed that there is a *base year* defined for this index and that the value of the index at some other date is known (or estimated), then the relative change in price of an item compared to its cost during the base can be predicted. Alternatively, if the cost at any year in the record is known, the cost at any other time can be estimated by setting up a ratio of the index values. In equation form this is

$$\begin{matrix} \text{Cost at} \\ \text{date of} \\ \text{interest} \end{matrix} = \text{Reference cost} \left[\frac{\textit{Index at date of interest}}{\textit{Index at date of reference cost}} \right] \quad (6.2)$$

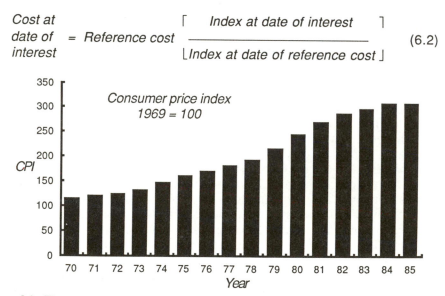

Figure 6.3 The variation of the consumer price index. This index represents a composite of items that consumers purchase (USDC, 1986).

It turns out there are several indices that reflect the many varied sectors of our broad economy. Most of people are familiar with the consumer price index (actually there are several consumer price indices, but here only a single, composite value for each year will be desired). This is a good indicator of inflation/deflation as it generally affects the wage earner. Quotations of the value of the CPI are given regularly on the evening news; and this is generated from food, housing, and other consumer product prices. A plot of the CPI is shown in Figure 6.3.

Two indices pertinent to plant design will be noted here. First is the Marshall and Swift (M&S) Equipment Cost Index. This one is actually a dual tabulation of both "all-industry" and "process-industry" components. The all-industry portion is an arithmetic average of 47 different components related to commercial, industrial, and housing equipment. The process-industry portion is a weighted average of a subset of eight of these with heavy emphasis on chemical and petroleum aspects. Both of these are defined relative to a base of 100 in the year 1926. Reflected in both portions of the index are costs of major equipment and machinery as well as costs for installation and necessary minor equipment required for the installation of the major equipment. Values of this index are published regularly in Chemical Engineering magazine. A tabulation for recent years is given in Table 6.1, and these values are shown in Figure 6.4.

Table 6.1
Various Cost Indexes, 1970-1984

YEAR	M&S[a]	EN-R[b]	CPI[c]
1970	303	1385	116.3
1971	321	1581	121.3
1972	332	1753	125.3
1973	344	1895	133.1
1974	398	2020	147.7
1975	444	2212	161.2
1976	472	2401	170.5
1977	505	2577	181.5
1978	545	2776	195.4
1979	599	3003	217.4
1980	660	3237	246.8
1981	721	3537	272.4
1982	746	3825	289.1
1983	761	4055	298.4
1984	780	4146	310.7
1985	790	4182	309.5

a. M&S equipment cost index, annual index (1926=100). Chemical Engineering, January 20, 1986, p. 7. b. EN-R construction cost index, annual average. Engineering News-Record, March 20, 1986, p. 107. c. Consumer price index, all items. STATISTICAL ABSTRACT OF THE UNITED STATES 1986, 106th EDITION. U.S. Department of Commerce, 1986, p. 472.

Another index commonly used for updating costs is the Engineering News-Record construction index. A composite weighting of the costs of common construction materials (structural steel, lumber, and cement) and common labor are taken to generate this index. There are three bases given: 100 in 1913, 100 in 1949, and 100 in 1967. An update of this index is given weekly in the <u>Engineering News-Record</u>, and a yearly summary is given each March.

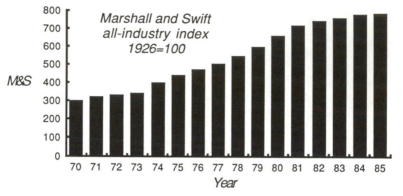

Figure 6.4 The variation of the Marshall & Swift (M&S) all-industry annual average index of equipment prices. A base of 100 was taken in 1926. Plotted from data tabulated in the <u>Chemical Engineering</u> magazine, January 20, 1986 and earlier editions.

A comparison of the three indicators discussed above is of interest. Since they each have a different base, and since it will be shown that the relative change in an index will be what is important for calculations, the percentage change by year for all indices is shown in Figure 6.5. Also shown there is the change in the Gross National Product (GNP) implicit price deflator. While the four variations are not identical, they do reflect common trends.

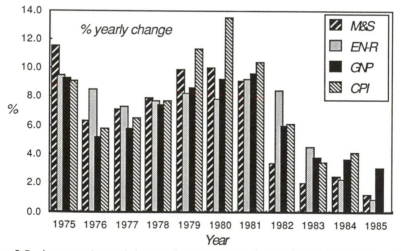

Figure 6.5 A comparison of the yearly percentage change in the Marshall and Swift equipment price index (M&S) and the Engineering News-Record construction index (EN-R) with the Gross National Product (GNP) implicit price deflator and the Consumer Price Index (CPI). The values shown are for the change from the preceding year to the listed year. Data taken from USDC (1986) for the CPI, CEA (1986) for the GNP, <u>Chemical Engineering</u>, January 20, 1986 for the M&S, and <u>Engineering News-Record</u>, March 20,1986. Note that the CPI change for 1985 is a very small, negative number.

Other indices are tabulated that reflect the costs of such specific areas as chemical plant costs, refinery construction costs, and other types of plant costs, both domestic and worldwide. Many of these indices are reviewed and tabulated in several texts. (See, e.g., Woods, 1975; Peters and Timmerhaus, 1980.) Different indices might yield different results, which can be disconcerting. However, the index is only as good as the data incorporated into it. Also, the data may reflect differing conditions. It is well known that bids on construction projects vary widely depending upon the economic situation that exists. Hence, if there is a period when a contracting concern is short of business, that company may artificially underbid a particular job to make sure it has work for its employees. This might show up as a decrease in the overall index, even though it might not represent a trend.

A single cost index is preferred here for consistency and clarity. Since a major portion of the cost of a thermal system often is due to equipment and a lesser portion consists of the actual construction process, a cost index that emphasizes the equipment costs is preferred. Another desired characteristic of a cost index to be used here is that tabulations of the history of the index be readily available. The Marshall and Swift index meets these requirements. M&S values are readily available because they are tabulated regularly in Chemical Engineering, for example. Use of the M&S index is patterned after some previous plant cost tabulations. See Peters and Timmerhaus (1980) and the various references by Woods and co-workers.

All costs listed in Table D.3 have been adjusted to an M&S index value of 800. Correct these values as appropriate for other applications.

Engineers who have reason to believe another index would suit their purposes better than the M&S should use the better one. The translation of the costs tabulated here and in other sources can be easily accomplished by the ratio approach with the appropriate indices. Keep in mind, however, that much of the cost information given in Appendix D has been updated using the M&S index.

One final point should be noted about inflation/deflation factors. As discussed in Chapter 5, there can be situations where a number of factors with differing rates can be present at the same time. A graphic example of this can be seen by comparing energy cost inflation to general inflation rates during the period 1973-1980, approximately. During that time, certain energy costs were inflating at nearly twice the rate of the rest of the economy. This can be very important when the two rates affect two aspects of the same design. It should be noted that attempts to project the future costs of energy are fraught with potentially large errors. However, some authors have presented such estimates (e.g., Klepper, 1980).

6.2.3 The Factor Method

Now that up-to-date costs of the major equipment as a function of the equipment type and size are available, is there any way that the total cost of the project can be estimated without going through a complete plant cost evaluation? The answer is "yes," and the accuracy of the estimate depends very much on the level of effort you wish to put into the evaluation process. One of the simpler techniques that may have a low value of accuracy is termed the *factor method.*

This technique uses the costs of major components of equipment to calculate the full plant cost. Expressed in equation form,

$$C = [\sum_{i=1}^{n} F_i C_i][F_I + 1]$$
(6.3)

Here

C = total cost of project,

C_i = costs of the major pieces of equipment,

F_i = related cost factors for associated building or other nonequipment capital facilities,

F_I = indirect (or overhead) cost factor, and

n = number of pieces of major equipment.

This technique is sometimes referred to as the *ratio, percentage,* or *parameter method.*

When a single item dominates the equipment list, such as the collector field in a solar power plant, then the factor method can often be written in simplified form:

$$C = FC_e$$
(6.4)

Here C_e is the cost of the single major item and F is the adjustment factor to account for all auxiliaries and overhead costs. This is sometimes called the *unit cost-estimating method.*

The price of the project will be affected by its construction location, elapsed time allowed for construction, and, of course, the precise details of the plant design. It is usually the latter factors (which can be very large in number) that are handled by the techniques already discussed.

A word of caution is in order. The type of plant can have a great effect on the price. For example, chemical plants and ocean thermal energy conversion plants could have basic differences in costing for essentially the same equipment. Since this is an effect that influences the price of very specialized plants, it cannot be explored here in any detailed way.

(For further information on the factor methods, see Guthrie, 1969; Guthrie, 1974; Woods, 1975; Peters and Timmerhaus, 1980; Vatavuk and Neveril, 1980b; Lunde, 1984; Ward, 1984.)

6.3 USING THE TABLES—EXAMPLES

Example 1

Required--
  Estimate the approximate price for a water pump that will flow 500 gpm of water through a 10-psi head. No motor is to be included in the price.

Solution--
  Since the head is not too high and the flow is fairly large, it is probable that a centrifugal pump will be appropriate for this application. With no other information given, assume that this is the case.
  The appropriate category from Table D.3 is listed below for convenience.

PUMP Centrifugal, horizontal, 50 ft	.26	2	10	.2↔16 kW
head, ci, radial flow, no motor. FOB.	.43	5.3	100	16↔400 kW
(Others given in reference.) {1}	.34	3.2	.5	.05↔30 m³/min

Remember that the first column of numbers refers to values for m in Equation 6.1, the second column is C_r in thousands of dollars, the third column is S_r, and the fourth column is the $(S_{min} \leftrightarrow S_{max})$ range. The units for S_r are the same as for $(S_{min} \leftrightarrow S_{max})$.
  The first step is to see if the specifications fit this application. Everything in the physical description appears to be appropriate. Checking the 50-ft head aspect ("ft head" always refers to "ft of water" unless stated otherwise), this application is equivalent to a pressure head of about 23 ft: this is smaller than the tabulated maximum and thus meets the requirement. The listing is for cast iron, and since there is no information to the contrary, this will be taken to be a satisfactory construction.
  Next, determine the line of numbers, if any, that will be appropriate. The first two lines are in terms of power. There may be some times when pump power is calculated in a simulation, and with an assumption about pump efficiency, this sizing factor could be used here. Here, however, make the calculation in terms of flow rate instead. Converting gpm, find the specified flow to be 1.89 m³/min. This value falls within the minimum and maximum values of the range, so this tabulation can be used. The cost is then estimated with Equation 6.1.

$$C = C_r (S/S_r)^m = \$3200\,(1.89/0.5)^{0.34} \approx \$5030 \qquad \text{Solution}$$

As noted in the table, this is a FOB cost for a time when the M&S index value is 800. Although more significant digits could be shown in the final answer, they are not warranted with the approximations inherent in this approach.

Example 2

Required--
   Estimate the price of a forced-draft cooling tower that will cool 25,000 gpm of water for the following conditions: wet-bulb temperature of 75°F, an approach temperature difference of 20°F, and a cooling range of 15°F.

Solution--
   Cooling towers are listed under the Heat Exchangers heading in Table D.3. Since this category is quite lengthy, it will not, as was done in Example 1, be repeated here. Note that size listings are either in terms of water flow rate or cooling capacity. Either one can be calculated from the given information, but the listings for flow will yield the shortest calculation procedure. First convert the flow to SI units.

$$25000 \text{ gpm} = 94.6 \text{ m}^3/\text{min} = 1.58 \text{ m}^3/\text{s}$$

This value fits within the second line of values. Again using Equation 6.1:

$$C = C_r \, (S / S_r)^{\,m} = \$560,000 \, (94.6 / 100)^{0.64} \approx \$540,000$$

Reading further in Table D.3, note that implicit in this cost are certain values for wet-bulb temperature, approach temperature difference, and cooling range temperature difference. Other values have correction factors associated with them. The correction for approach temperature difference is 0.49 because the value here is 20°F (= 11°C), and the correction factor is found in the first column. A wet-bulb temperature here of 75°F is the same as that used in the initial listings, so no correction for this is needed. Finally, our value of cooling range is 15°F (= 8.3°C), which yields a value of 1.19 interpolated from the table. Thus, the estimated cost for the given conditions is:

Cooling tower: $C = (1)(0.49) \, (1.19) \, (\$540,000) \approx \$320,000$

This price represents an installed unit except for the foundations, water pumps, and distribution pipes. Another category lists numbers that can be used for estimating the cost of the distribution piping, although this will be so site specific that the value will have a much higher potential error than the value estimated for the cooling tower.

Water distribution: $C = \$160,000 \, (1.58 / 1.0)^{0.7} = \$220,000$

   Note that there is a second heading for Cooling Towers in Table D.3 that makes reference to Vatavuk and Neveril (1981b). To find a second estimate, refer to that paper. Checking their Figure 3 shows a more detailed functional form than that used here. Using their form, the cost is found to be

$$C = (C_C - 34{,}500) F_1 F_2 + 34{,}500 =$$
$$(180{,}000 - 34{,}500) (1.27) (0.50) + 34{,}500 \approx \$127{,}000$$

According to Vatavuk and Neveril, "included...is the price of the tower, fans, pumps, and motors, and the expense of their installation but not the price of the basin." At first glance, this number appears to be <u>considerably</u> smaller than the $320,000 estimated earlier. It might be that the data in the paper and in Table D.3 here do not correspond to the same year. As should be the case whenever using historical data, be concerned with the year that the data were correlated. In periods of high inflation, old data really needs to be updated! Checking the Vatavuk and Neveril paper, they state that "except when specifically noted otherwise, all costs have been updated to December 1977." Updating this is the topic of Example 3.

Example 3

Required--
   A value of $127,000 was found for a cooling tower using 1977 data from a paper in <u>Chemical Engineering</u>. How should this be adjusted to compare to a corresponding calculation using data in Table D.3?

Solution--
   Table D.3 has all values adjusted to an M&S equipment index equal to 800. Find the corresponding value of the M&S index for 1977. From Table 6.1, the value is 505. Using Equation 6.1, the updated cost is

$$C = \$127{,}000 (800 / 505) \approx \$200{,}000$$

There is still a discrepancy between the two predictions made in Example 2, but the difference is smaller than it was before. To rationalize the differences would take an in-depth look at the cost data used in both correlations. Since the Vatavuk and Neveril estimation approach is both newer and more comprehensive than the one in Table D.3, it may yield better accuracy. However, the magnitude of the variations shown here is typical of many calculations.

6.4 PREDICTION OF ENERGY COSTS

   In contrast to the space given to <u>historical</u> prices for capital equipment in the previous sections of this chapter, the focus on energy costs is into the <u>future</u>. Capital equipment prices are influenced primarily by inflation/deflation effects. Energy costs can be influenced by the same factors and a great many more. An example of this can be seen by

131

examining variations of the "all-items" consumer price index with the special category CPI for gas and electricity. These are clearly contrasted in Figure 6.6. Of course, the CPI may not be the precise measure of inflation needed for industrial plant design (see the comparisons made earlier in this chapter), but similar trends would be seen in similar comparisons of industrially based indicators.

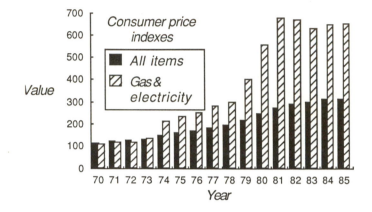

Figure 6.6 Comparison of the all-items and gas & electricity CPIs for the years 1970-1985. A very marked difference in inflation rates is shown (USDC, 1986).

While current energy costs undoubtedly have a major impact on existing plant operations, new plant designs require that a forecast of the price of energy be considered. Current energy costs are easily found from the local supplier(s). Accurate forecasts of future energy costs are always more difficult to come by.

Future costs of energy can be influenced by a number of variables, not the least of which is the political climate both here and abroad. Few people anticipated the rapid increase in the cost of oil that was experienced during the early and mid-1970s. Most people were also surprised at the counteracting decrease of oil prices in the early and mid-1980s. Many of these variations were due to political factors acting upon the energy market. Similar trends were seen in the price of natural gas. Electrical energy prices were influenced during the latter part of the 1970s by demand. Many utilities were faced with a need to increase capacity, and they found that this was a very expensive undertaking for a variety of reasons. Higher electrical prices were generally the result. An interconnecting effect of energy prices is also found. When the price of one major source changes significantly up or down, a ripple effect to some degree is in most of the other sources. This can be due to direct impacts (such as the substitution of coal for natural gas when the latter was prohibited for heavy industrial use) or less obvious market effects encouraging the substitution of a cheaper source for a more expensive one.

There is most certainly an element of crystal ball gazing, or just plain guessing, when trying to predict the future prices of energy. However, some ongoing efforts attempt to bring into these considerations various types of anticipated market influences in a quantitative way. One such effort is that being pursued commercially by Data Resources, Inc. (DRI, 1985). Like many of the related efforts, this one attempts to analyze recent and current energy use practices and prices and then incorporates these aspects with predicted market forces in the years ahead.

The results of an autumn 1985 Data Resources, Inc., tabulation of histories and predictions are shown for several key sources in Table 6.2. A plot of the various yearly percentage changes is shown in Figure 6.7.

Table 6.2

Average Energy Price History and Forecast (DRI, 1985)[a]

Year	Coal		Electricity		Natural Gas		Oil	
	$/MMBtu	% Change	$/kWh	% Change	$/therm	% Change	$/MMBtu	% Change
1981	1.806	12.0	0.040	17.1	0.308	19.7	5.234	37.4
1982	1.964	8.8	0.047	15.6	0.384	24.7	4.684	-10.5
1983	1.961	-0.2	0.047	0.5	0.417	8.4	4.511	-3.7
1984	2.121	8.2	0.049	5.0	0.409	-1.8	4.578	1.5
1985	2.098	-1.1	0.049	-0.5	0.374	-8.5	4.100	-10.4
1986	2.103	0.3	0.049	0.8	0.332	-11.2	3.641	-11.2
1987	2.136	1.6	0.050	1.3	0.308	-7.2	3.376	-7.3
1988	2.203	3.1	0.051	2.8	0.308	0.0	3.398	0.7
1989	2.290	3.9	0.053	2.6	0.319	3.3	3.512	3.4
1990	2.416	5.5	0.054	3.3	0.330	3.6	3.655	4.1
1991	2.573	6.5	0.057	4.5	0.354	7.2	3.880	6.1
1992	2.781	8.1	0.060	4.6	0.377	6.5	4.088	5.4
1993	3.001	7.9	0.063	5.1	0.406	7.8	4.371	6.9
1994	3.260	8.6	0.066	5.5	0.441	8.7	4.714	7.8
1995	3.525	8.1	0.070	5.8	0.483	9.4	5.124	8.7

a. "% Change" denotes yearly value.

Keep in mind that both Table 6.2 and Figure 6.7 represent a great simplification to the real situation even if the predicted trends turn out to be totally accurate. Energy prices vary considerably with both the geographical location of the energy use as well as the industrial sector where utilization occurs. These factors will influence also the relative changes in the cost of energy. For simplicity, it is recommended that if more detailed information is not available for preliminary design purposes, the current, local energy price situation for the type of industry should be used with the relative trends shown in Table 6.2. More detailed predictions are available from a variety of sources, including DRI.

133

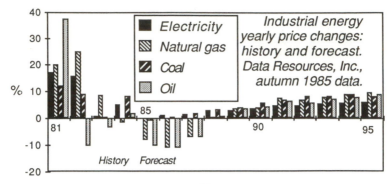

Figure 6.7 History and forecast of yearly percentage price changes of four prime energy sources for the period 1981-1995, current during the autumn of 1985. From Data Resources, Inc., information (DRI, 1985).

REFERENCES

CEA, 1986, ECONOMIC REPORT OF THE PRESIDENT, U.S. Government Printing Office, Washington, DC, February.

DRI, 1985, Data Resources, Incorporated, Energy Service, Autumn 1985 Forecast, courtesy of DRI, San Francisco, CA, October 30.

Guthrie, K., 1969, "Data and Techniques for Preliminary Capital Cost Estimating," Chemical Engineering, March 24, pp. 114-142.

Guthrie, K., 1974, PROCESS PLANT ESTIMATING EVALUATION AND CONTROL, Craftsman Book Company, Solano Beach, California.

Humphreys, K., and S. Katell, 1981, BASIC COST ENGINEERING, Marcel Dekker, New York.

Lunde, K., 1984, "Capacity Exponents for Field Construction Costs," Chemical Engineering, March 5, pp. 71-74.

Means, 1985, MEANS MECHANICAL COST DATA, 1986, 9TH ANNUAL EDITION, R. S. Means Co., Kingston, MA.

Peters, M., and K. Timmerhaus, 1980, PLANT DESIGN AND ECONOMICS FOR CHEMICAL ENGINEERS, THIRD EDITION, McGraw-Hill, New York.

USDC, 1985, STATISTICAL ABSTRACT OF THE UNITED STATES, 1985, 105th EDITION, U. S. Department of Commerce, Section 16.

Vatavuk, W., and R. Neveril, 1980a, "Estimating Costs of Air-Pollution Control Systems, I. Parameters for Sizing Systems," Chemical Engineering, October 6, pp. 165-168.

Vatavuk, W., and R. Neveril, 1980b,----"II. Factors for Estimating Capital and Operating Costs," Chemical Engineering, November 3, pp. 157-162.

Vatavuk, W., and R. Neveril, 1980c,----"III. Estimating the Size and Cost of Pollutant Capture Hoods," Chemical Engineering, December 1, pp. 111-115.

Vatavuk, W., and R. Neveril, 1980d,----"IV. Estimating the Size and Cost of Ductwork," Chemical Engineering, December 29, pp. 71-73.

Vatavuk, W., and R. Neveril, 1981a,----"V. Estimating the Size and Cost of

Gas Conditioners," Chemical Engineering, January 26, pp. 127-132.

Vatavuk, W., and R. Neveril, 1981b,----"VI. Estimating Cost of Dust-Removal and Water- Handling Equipment," Chemical Engineering, March 23, pp. 223-228.

Vatavuk, W., and R. Neveril, 1981c,----"VII. Estimating Costs of Fans and Accessories," Chemical Engineering, May 18, pp. 171-177.

Vatavuk, W., and R. Neveril, 1981d,----"VIII. Estimating Cost of Exhaust Stacks," Chemical Engineering, June 15, pp. 129-130.

Vatavuk, W., and R. Neveril, 1981e,----"IX. Costs of Electrostatic Precipitators," Chemical Engineering, September 7, pp. 139-140.

Vatavuk, W., and R. Neveril, 1981f,----"X. Estimating Size and Cost of Venturi Scrubbers," Chemical Engineering, November 3, pp. 93-96.

Vatavuk, W., and R. Neveril, 1982a,----"XI. Estimate the Size and Cost of Baghouses," Chemical Engineering, March 22, pp. 153-158.

Vatavuk, W., and R. Neveril, 1982b,----"XII. Estimate the Size and Cost of Incinerators," Chemical Engineering, July 12, pp. 129-132.

Vatavuk, W., and R. Neveril, 1982c,----"XIII. Costs of Gas Absorbers," Chemical Engineering, October 4, pp. 135-136.

Vatavuk, W., and R. Neveril, 1983a, ----"XIV. Costs of Carbon Adsorbers," Chemical Engineering, January 24, p. 131.

Vatavuk, W., and R. Neveril, 1983b, ----"XV. Costs of Flares," Chemical Engineering, February 21, pp. 89-90.

Vatavuk, W., and R. Neveril, 1983c, ----"XVI. Refrigeration Systems," Chemical Engineering, May 16, pp. 95-98.

Vatavuk, W., and R. Neveril, 1984a, ----"XVII. Particle Emissions Control," Chemical Engineering, April 2, pp. 97-99.

Vatavuk, W., and R. Neveril, 1984b, ----"XVIII. Gaseous Emissions Control," Chemical Engineering, April 30, pp. 95-98.

Ward, T., 1984, "Predesign Estimating of Plant Capital Costs," Chemical Engineering, September 17, pp. 121-124.

Woods, D., 1975, FINANCIAL DECISION MAKING IN THE PROCESS INDUSTRY, Prentice-Hall, Englewood Cliffs, NJ.

Woods, D., S. Anderson, and S. Norman, 1976, "Evaluation of Capital Cost Data: Heat Exchangers," The Canadian Journal of Chemical Engineering, 54, December, pp. 469-488.

Woods, D., S. Anderson, and S. Norman, 1978, "Evaluation of Capital Cost Data: Gas Moving Equipment," The Canadian Journal of Chemical Engineering, 56, August, pp. 413-435.

Woods, D., S. Anderson, and S. Norman, 1979a, "Evaluation of Capital Cost Data: Liquid Moving Equipment," The Canadian Journal of Chemical Engineering, 57, August, pp. 385-408.

Woods, D., S. Anderson, and S. Norman, 1979b, "Evaluation of Capital Cost Data: Offsite Utilities (Supply)," The Canadian Journal of Chemical Engineering, 57, October, pp. 533-565.

Yeoman, J., 1978, "Wind Turbines," ANL/CES/TE 78-9, December. (Available from the National Technical Information Service, U.S. Department of Commerce, 5285 Port Royal Road, Springfield, Virginia 22161.)

PROBLEMS

6.1 Determine the cost of centrifugal pumps by contacting vendors. Consider only pumps with carbon steel construction and without any auxiliary accessories such as drive motors, mounting plates, flanges, and so on. **(a)** Tabulate a description of the pump, including the manufacturer's name, the required drive horsepower, some indication of the pump flow versus head characteristics, a pump efficiency, and the cost. **(b)** Give a graphical representation of this cost information. On the same graph, show an updated form of the historical data tabulated in this chapter. **(c)** Develop a mathematical curve fit of your data and show this function on the same plot as your data. Attempt to describe more than one independent variable in the function.

6.2 Collect and analyze the cost data for gas-fired water heaters and present the results in an appropriate manner similar to the steps indicated in Problem 6.1. Compare these to data from Means (1985).

6.3 It is desired to store water in an above-ground tank. Show the cost of the tank as a function of the size of the tank (between 1000 and 100,000 gal) on a graph for plastic, steel, and concrete construction. Assume that the top of the water is at atmospheric pressure. Is there a preferable shape for each of these types of construction?

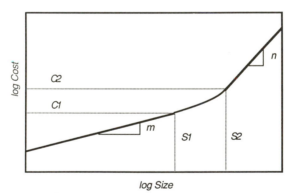

Figure 6.8 Hypothetical cost function to be used in solving Problem 6.4.

6.4 The cost versus capacity of some items is noted to vary as shown in Figure 6.8. Determine a mathematical function to describe this behavior. Choose some numbers for a plot and show how you would solve for the constants and what your representation of the plot gives. Assume that the plot shown is on log-log coordinates and that the cost representation is a straight line below *S1* with slope *m* and is a straight line above *S2* with slope *n*. The solution you find should be valid for *m* both greater than *n* and less than *n*. You may use other characteristic points on the curve in defining your function, but be sure to define any other point you use.

6.5 Contact your local utility or public utilities commission (all states have some organization of this sort that regulates the public utilities, although its name may vary from state to state) to determine the cost of electricity. **(a)** Show how the cost of electricity varies with the amount used on a monthly basis. Plot this variation in a suitable manner. **(b)** Plot the variation of the cost of electricity since 1970. If there has been different total changes for the various rate schedules, choose only one to plot that you deem to be a schedule that might be applicable to a medium-sized manufacturing industry. Compare the variation in electrical rates with the variation in the Marshall and Swift equipment index variation and the electrical energy cost history/prediction(s) for the pertinent years shown in this chapter.

6.6 Compare the base capital costs for the following three options for power generation over the rar.ge of outputs from 100 kW to 10 MW: **(a)** diesel engine, **(b)** simple gas turbine, and **(c)** simple Rankine cycle (assuming this is made up of a boiler, turbine, condenser, and pump). Include any heat removal equipment that may be necessary, but neglect the cost of the electrical generation and transmission equipment costs.

6.7 With the foundation of Problem 6.6, now focus on the same three power plants in a 10,000-HP size. Determine the relative cost of fuels that would make the 15-year life cycle costs the same for the three options. Use the following assumptions. The load factor is 0.6 over the total 15-year period for all engines. Take the overall thermal efficiencies as follows: gas turbine =18%; diesel=26%; Rankine=38%. Operation and maintenance amounts to 5% of the purchase price on a yearly basis. There is no salvage value after 15 years. Total installed costs of each configuration are twice the equipment costs.

6.8 Solicit information from local vendors to determine the cost of at least three common building insulations. Present this data in a manner appropriate for comparison. While the simplest representation may be $/*R* value, you should also consider giving some insights into installation costs (which undoubtedly are not linear with *R*) and the volume required for given *R*. (If the basic wall constuction must be made thicker than normal to incorporate a given *R* rating, this certainly has cost implications. Can you show an effect of this sort into your cost correlation?)

6.9 Compare the total inflation reflected by the CPI, M&S, and EN-R for the time periods 1973-1976; 1973-1978; 1973-1980; 1973-1982; 1973-1984; and 1977-1984.

6.10 An internal combustion engine is to be mounted in a laboratory. Determine a size and cost of a shell-and-tube heat exchanger to accomplish this task. A cooling water supply of 10 gpm (if this much is needed, use it, but remember also that water costs) and 70°F is available. The engine has a rated power output of 200 HP and a thermal efficiency of 28%. A 50% ethlylene glycol solution is circulated through the engine, and the temperature at which the coolant returns to the engine should never be higher than 100°F at rated conditions.

6.11 Estimate the 1980 prices of the following heat exchangers for the range of areas of 10 $\leftrightarrow$ 1000 ft^2: carbon steel shell and tube; 304 stainless steel shell and tube; carbon steel double pipe; and 304 stainless steel plate. Show these on a graph. Assume that all have low-pressure duties. Do not show any correlation extending beyond its given range.

6.12 Correlate the costs for butterfly valves, 175 lb rating, full port, threaded, and constructed from type 316 stainless steel. Attempt to do this in a form similar to that used in Table D.3 of the Appendix. You may wish to contact a vendor for this information or consult any pertinent listings in the literature (e.g., Means, 1985).

6.13 Compare the prices predicted in Table D.3 for air-to-air heat pumps with any recent data. List any major factors that influence costs other than the correlating variable used in Table D.3.

6.14 The following pairs of data are tabulated for reciprocating hermetic refrigeration compressors (bare equipment costs) as a function of the cooling rating of the vapor compression refrigeration system (ratings are ARI standard 515 group IV using R-22) (Means, 1985):

2 ton-$1950. 15 ton-$4225. 20 ton-$5525. 40 ton-$6110. 50 ton-$13,300. 75 ton-$16,370. 130 ton-$19,500.

(a) Correlate this data in the same context and form used in Table D.3 of the Appendix. List the maximum error.

(b) If there is a functional form that fits the data better than the simple power function form used in Equation 6.1, determine it. Cite the maximum error.

6.15 The costs of high capacity (at 0.4 in. SP) evaporative coolers that operate on 230/460 V power have been tabulated (Means, 1985) and are listed below. Determine a correlation in the form used in Table D.3 to represent these costs functionally. What is the maximum error?

12,100 cfm, 2 HP, $3600. 14,000 cfm, 3 HP, $3700. 17,100 cfm, 5 HP, $3850. 22,500 cfm, 7.5 HP, $5500. 25,100 cfm, 10 HP, $5850. 29,500 cfm, 15 HP, $6150.

6.16 Means (1985) lists costs for reciprocating, water-cooled, heavy-duty, slow-speed lubricated air compressors with a 100-psi rating as follows (bare machine costs are shown):

73 cfm by 20 HP motor= $10,900; 100 cfm by 25 HP motor= $11,450; 132 cfm by 30 HP motor= $13,900; 185 cfm by 40 HP motor= $16,500; 210 cfm by 50 HP motor= $18,550; 330 cfm by 75 HP motor= $25,850; 557 cfm by 125 HP motor= $36,950;

Correlate this data in a form like that used in Equation 6.1 and Table D.3. Indicate the maximum error between the data and your correlation. Also determine how your correlation compares to the most appropriate correlation given in Table D.3.

6.17 By referring to the appropriate literature, give tabulations similar to those shown in Table 6.1 but for the most current five-year period. That is, update the information given in Table 6.1 for the past five years.

6.18 What is an appropriate numerical value to use in the brackets in Equation 6.2 to update the information given in Table D.3 to the present? Note that you will need to consult the literature for the most current index data.

CHAPTER 7
AVAILABILITY ANALYSIS

7.1 INTRODUCTION

Applications from the field of thermodynamics are critical to the design of thermal systems. Any component in a steam power plant, for example, is usually analyzed during the design process with at least the First Law of Thermodynamics. Chances are very great that Second Law concepts also will have to be used in determining the overall performance. These kinds of calculations are the focus of the basic courses in thermodynamics. It is assumed that the reader has already acquired a background knowledge in this area.

Some applications of the Second Law are extremely valuable from a design standpoint but are covered very briefly, or not at all, in introductory courses. These involve the concept of availability analysis. These are also known as "available energy," "exergy," "essergy," or "lost work" analyses, although there are some minor distinctions between some of these topics. Theoretically, the concept of availability deals with the quality of energy, using work (as compared to heat) as the highest possible measure of the output of the thermodynamic system. Even the relative performance of purely heat transfer devices can be assessed on a more general basis through the use of an availability analysis.

In recent years, availability analyses have been reported for a variety of plant design applications. The power of this technique has been demonstrated, particularly when the economic trade-offs between various forms of energy must be compared. For example, it is often the case that heat and work outputs must be evaluated and values assigned to each. It may also be the case that various forms of chemical energy must have a value assessed compared to heat input. These kinds of comparisons may not have a basis without Second Law ideas through the concept of availability.

In what follows, three aspects of availability analysis will be covered. First, some background and definitional development will be given in Section 7.2. In Section 7.3, applications to conceptual designs will be described, where the Second Law will be shown to give an overall assessment of processes and systems. Applications to, and evaluations of, the relative economic value of heat and work in systems such as those used for cogeneration will be described in the final section (7.4). Extensive references to published works are given to facilitate additional reading, if desired.

7.2 BACKGROUND

While the ideas that form the background for the concept of availability were formulated in the nineteenth century, actual development of the thoughts for engineering applications took place in the 1920s and 1930s (Darrieus, 1930; Keenan, 1932). Now the description of the basic definitions and simple examples are found in most textbooks on engineering thermodynamics. At least six books giving fundamental and applied treatments of the concept have appeared (Bruges, 1959; Gaggioli, 1980; Ahern, 1980; Bejan, 1982; Moran, 1982; Kotas, 1985). Some of the key definitions will be outlined here to give a basis for design quantities to be used later.

7.2.1 Availability Definitions

The classical example of the concept of availability can be examined by considering the transfer of a quantity of heat from a thermal reservoir at T_h to a device or reservoir at T_c, with $T_h > T_c$. This process is obviously irreversible, consisting of heat transfer through a finite temperature difference. If it is imagined now that the heat transfer from the high-temperature reservoir, Q_h, takes place through a Carnot engine, which, in turn, rejects heat to the low-temperature reservoir, then net work, W, conceivably can be produced. In fact, that amount of work is given by Equation 7.1.

$$W = Q_h (1 - T_c / T_h) \qquad (7.1)$$

The quantity in parentheses is recognized as simply the Carnot efficiency between the two given reservoirs. Hence, when the simple heat transfer (actually the heat loss) occurs, as it does regularly in nature, the production of some quantity of work is forgone. In simple terms, this *lost work* is the loss in availability that results. Note that, unlike energy, which is always conserved through its changes in form, availability can be dissipated. The cooling of a warm object in contact with the ambient is an example of this. See Figure 7.1. In any state between *i* and *f,* there is a certain amount of energy that is *available* and a certain amount that is *unavailable.* If the available energy transferred as heat is not used in a work process, it becomes *destroyed available energy,* or *lost work.* More discussion of Second Law analysis of heat transfer processes is given later in this chapter.

Drawing upon analogy to the possibility of producing work during a

141

heat transfer process, the *thermodynamic availability* is defined as the amount of work that can be produced when a substance at a given state reverts to the *dead state* while producing work with a Carnot cycle. This can apply to any type of process. The dead state is usually defined as atmospheric conditions, and the hypothetical Carnot cycle is assumed to be rejecting heat to the ambient temperature, T_o. While distinctions are rightfully made between closed and open system availability in all developments in the literature, we will be concerned here only with the appropriate quantities for open systems. The flow availability for steady state, steady flow, *b*, is defined in Equation 7.2, where terms accounting for nuclear and other forms of energy that could be important in some systems have been omitted.

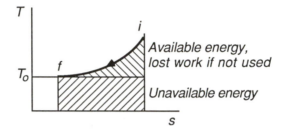

Figure 7.1 A consideration of the details of an object cooling from temperature T_i to the ambient temperature shows that part of the energy is *available* (i.e., it could be made into work with the use of a Carnot Cycle) and part of it is *unavailable*. If the available energy is not used to do work as the process proceeds, this energy becomes *lost work*.

$$b \equiv h - h_o - T_o(s - s_o) + g(z - z_o)$$

$$+ (V^2 - V_o^2)/2 + \sum_j n_j(\mu_j - \mu_{jo}) \qquad (7.2)$$

also $B = bm$ and, for steady state, steady flow, $\dot{B} = b\dot{m}$. Here *h* is the specific enthalpy of the substance, *T* is temperature, *s* is the specific entropy of the substance, *z* is the elevation, *g* is the local acceleration of gravity, *V* is the velocity, n_j is the number of moles of each specie in the substance, and μ_j is the chemical potential of each specie in the substance. The subscript *o* denotes environmental, or other dead state, conditions. Note that V_o will usually be taken as zero.

Evaluation of all but the final group of terms in Equation 7.2 involves simply heat capacity and physical properties of the substance of interest. The calculation of the terms involving the chemical potential can be somewhat more involved (Marin and Turegano, 1986). For pure condensed phases in a stable reference environment,

$$\sum n_j (\mu_j - \mu_{jo}) = 0$$

If the substance can be considered to be an ideal gas, then

$$\mu_j - \mu_{jo} = RT_o \ln (x_j / X_{j,o})$$

For all other situations, a chemical reaction between the compound of interest and the other components present is devised. In specific:

$$\kappa_j + v_1 R_1 + v_2 R_2 + \ldots + v_i R_i \Leftrightarrow v_{i+1} R_{i+1} + v_{i+2} R_{i+2} + \ldots + v_n R_n$$

here κ_j is the substance of interest. The chemical potential is found for the general case from the equation below.

$$\mu_j - \mu_{jo} = g_j(T_o, P_o, x_1, \ldots, x_n) + \sum_{k=1}^{i} v_k R_k - \sum_{k=i+1}^{n} v_k R_k$$

Here g_j denotes the Gibbs function. Calculations of the chemical potential for use in availability determinations are given in a number of references (Marin and Turegano, 1986). An excellent discussion of the chemical terms in the steady-flow equations has been given by Siemons (1986).

One general point of caution should be considered. Some controversy is found in the literature (de Nevers and Seader, 1980) regarding the definition of the dead state. This is not an easily defined quantity in some situations. In general, this may cause more problems in systems dealing with chemical reactions than in other areas of applied thermodynamics.

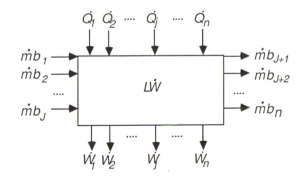

Figure 7.2 The availability analysis of the steady-state, steady-flow situation in a control volume is similar to the corresponding First Law analysis. There is always an accumulation of lost work within the control volume here, but there is no comparable situation in the First Law analysis.

Consider now the accounting of Second Law quantities in a general control volume. Refer to Figure 7.2. Using Equation 7.2, the availability, *b,* can be calculated for each stream flowing into and out of a thermodynamic device, and the net difference in all streams determined. This net flow of availability, when combined with other aspects, gives the result found in Equation 7.3, which is a special case for the steady-state, steady-flow situation.

$$\dot{W}_{max} = (\dot{W}_{cv} + L\dot{W}_{cv}) = \Sigma_j (\dot{m}\,b)_j + \Sigma_j \dot{Q}_j\,(1 - T_o/T_j) \qquad (7.3)$$

In Equation 7.3 the term $\dot{W}_{max}$ is the maximum possible rate that work can be performed by the control volume (*cv*); and *LW* is the rate of production of "lost work" within the control volume, a result of friction and other dissipative processes. *LW* is called the rate of production of "irreversibility" by some authors. Regardless of name, this quantity is always positive. $\dot{Q}_j$ is the rate of heat transfer from the *j*th isothermal section at temperature T_j of the control volume's surface. The temperature factor that multiplies the heat transfer terms is to account for the Carnot-type conversion that is part of the availability concept. $\dot{W}_{cv}$ shown in Equation 7.3 is simply the sum of all of the rates of actual work performed on the surroundings by the control volume. Note that the summation of the availability function assumes the flow into the control volume is positive, while the flow out is taken as a negative value in the sum. In all cases a dotted term denotes a rate per unit time.

Other authors have used a more direct statement of the Second Law to formulate similar kinds of concepts. (See, e.g., London, 1985.)

While the function defined in Equation 7.2 is appropriate for the determination of the availability of each stream of a simple compressible substance, there are several other kinds of situations and processes where availability must be determined. Consider, for example, the availability of thermal radiation. This is given by Equation 7.4 (Petela, 1964).

$$\dot{B} = \sigma A\,(3\,T^4 + T_o^4 - 4\,T_o\,T^3)/4 \qquad (7.4)$$

where σ is the Stefan-Boltzmann constant, *A* is the radiating area, *T* is the temperature of the radiation source, and T_o is the temperature of the surroundings.

Another quantity that is needed regularly for the analysis of thermal systems is the lost work resulting from fluid friction and its manifestation, pressure drop. Using a form of Equation 7.3 and the relationship *dh = T ds + v dP,* it can be shown that the rate of generation of lost work is given by Equation 7.5 (Moran, 1982).

144

$$\delta LW = \frac{T_O}{T} \left[\frac{8\dot{m}^3}{\pi^2 \rho^2 D^5} \right] f \, dL \tag{7.5}$$

where f is the friction factor defined by the relationship

$$(-dP/\rho) = V^2 f \, dL / 2D \tag{7.6}$$

and dL is the differential length of pipe whose diameter is D and ρ is the density of the fluid. As is the practice in some thermodynamics texts, the symbol d is used to define the *exact*, or point function, *differential* while δ is used to denote the *inexact*, or path function, *differential*.

The availability of various chemical fuels can be approximated by the heat of combustion (AIP, 1975). Availabilities of chemical fuels may, in fact, range up to about 110% of the heat of combustion. Complete tabulations for gaseous, liquid, and solid fuels (including solids containing small amounts of sulfur) have been given in general form in the literature. A corrected version in English (Rodriguez, 1980) of an earlier German work will prove quite valuable to designers. Related tabulations and calculations are also available (Moran, 1982; Shieh and Fan, 1982; Kenney, 1984).

For further information on the basic ideas of availability, check any of the various treatises on this topic. Included are Bruges (1959), Cozzi (1975), Ahern (1980), Moran (1982), and Kenney (1984) as well as many of the basic texts on thermodynamics like Van Wylen and Sonntag (1985).

Several tools have been introduced to be used in the analysis of simple processes. Now consider the definitions of some figures of merit so that various options can be compared in a meaningful way.

7.2.2 Second Law Efficiencies

Consider how process efficiencies are evaluated. To be really versatile, the comparison should be able to deal with two widely differing processes that accomplish the same goal. Some performance indicator is desired, whether it be for processes or groups of them in cycle form, which will allow a rational performance assessment.

It is of value to review briefly some concepts of efficiency. For processes, the appropriate efficiency is defined on the actual output of the device being compared to the ideal output of the device such that the ratio is less than 1. Consider the efficiency of an electric water heater. It is well known that the only significant "loss" of one of these devices is the so-called standby heat loss. For well-insulated heaters, this loss can be less than 10% of the total electrical input. Hence, it is not uncommon to see efficiency values stated as

145

$$\eta = \text{Useful heat out / Electrical input} \approx 90\%$$

This will be termed a *First Law Efficiency* here and is contrasted to a Second Law Efficiency to be defined below.

Thermal power cycles have a well-known relationship for efficiency; namely,

$$\eta_{cycle} = W_{net} / Q_{supplied} \tag{7.7}$$

where W_{net} is the net work produced by the cycle for each unit of energy supplied as heat, $Q_{supplied}$. In the present context, this relation will also be termed a *First Law Efficiency* because heat and work are compared directly.

A Second Law Efficiency will be one that will make a comparison between availabilities of actual devices and their ideal counterparts. This is an imprecise statement. The reason for the imprecision is that Second Law Efficiencies are less obvious in their definitions, being, in general, dependent upon the particular process or cycle of interest.

Although there are fundamental similarities, consider some of the Second Law Efficiency definitions given in previously reported work. One definition of a Second Law Efficiency, denoted here as ε_1, is the ratio of the output of availability divided by the availability input (Moran, 1982).

$$\varepsilon_1 = \frac{\text{Availability out in product}}{\text{Availability in}} \tag{7.8}$$

Allowing that there can be a variety of specific applications, Moran gives an efficiency for electrical power production as

$$\varepsilon_1 = W_{net} / B_f \tag{7.9}$$

where B_f is the availability of the fuel. Here the Second Law Efficiency is not too different from the First Law Efficiency as defined in Equation 7.7. Using the same general definition for a process application, Moran considers an electrical water-heating situation. The appropriate equation is given by Equation 7.10.

$$\varepsilon_1 = \Delta B / W \tag{7.10}$$

In this expression, W denotes the electrical input and ΔB is the availability gained by the water, found using Equation 7.2. An expression can be developed between the Second Law Efficiency, ε_1, and the First Law Efficiency, η, showing for typical operating conditions that $\varepsilon_1 \approx \eta / 20$.

Moran then examines a variety of cycles and other kinds of processes, including devices with only heat interactions, and shows that this general definition is sufficient. Other authors (e.g., Hevert and Hevert, 1980) have pointed out that this definition does not show how much availability has been wasted in relation to the theoretical maximum amount of work that could have been produced. However, the similarity between First and Second Law Efficiency values for work processes and cycles has appeal to many people.

Others (AIP, 1975; Kreith and Kreider, 1978) define another Second Law Efficiency; namely,

$$\varepsilon = B_{min} / B_{actual} \qquad (7.11)$$

where B_{min} is the minimum amount of availability required to accomplish a specific process, and B_{actual} is the actual amount consumed. This definition yields an effective way of visualizing many processes, as shown in Tables 7.1 and 7.2. The Second Law Efficiency of virtually any kind of thermodynamic process can be readily evaluated from these tabulations.

Table 7.1
Available Work Provided by Sources and Needed by End Uses[a,b]

		Fuel with Heat of Combustion $|\Delta H|$	Heat Q_1 from Hot Reservoir at T_1
SOURCE	Work, W_{in}		
	(e.g., water power, wind power, raised weight, electricity)	(e.g., coal, oil, gas)	(e.g., geothermal source solar collector source)
	$B = W_{in}$	$B \approx |\Delta H|^c$ (usually within 10%)	$B = Q_1(1 - T_0/T_1)$
END USE	Work, W_{out}	Heat Q_2 added to a hot reservoir at T_2	Heat Q_3 extracted from cool reservoir at T_3
	(e.g., turning shafts, pumping fluids) $B_{min} = W_{out}$	(e.g., space heating, cooking, baking, drying) $B_{min} = Q_2(1 - T_0/T_2)$	(e.g., refrigerating, air conditioning) $B_{min} = Q_3[(T_0/T_3) - 1]$

a. Source: AIP, 1975. Used with permisson. b. *Note:* T_1 (hot) > T_2 (warm) > T_0 (ambient) > T_3 (cool). c. See discussion earlier in this section.

There is no general rule about whether or not an efficiency is actually needed in a given analysis, or even which form to use. Sometimes the parameter to evaluate is obvious for a given process. While an efficiency could be computed, some other criteria may be used instead. Implicitly, however, an efficiency is calculated; and the various definitions given above should be kept in mind.

7.3 PROCESS APPLICATIONS

7.3.1 Introduction

In this section, some typical processes and components that might be encountered in the analysis of thermal systems are considered. A few examples are chosen to give the reader insight into how the application is handled, giving enough background to facilitate work in the future on these or other components. Special emphasis is given to the very important topic of heat exchange.

Table 7.2
First Law and Second Law Efficiencies for Single-Source, Single-Output Devices[a]

End Use	Source Work W_{in}	Fuel Heat of Combustion $|\Delta H|$ Available Energy B	Heat Q_1 from Hot Reservoir at T_1
Work W_{out}	1 $\eta = W_{out}/W_{in}$ $\varepsilon = \eta$ (e.g., elect motor)	2 $\eta = W_{out}/|\Delta H|$ $\varepsilon = W_{out}/B\ (\approx \eta)$ (e.g., power plant)	3 $\eta = W_{out}/Q_1$ $\varepsilon = \eta\ /[1-T_0/T_1]$ (e.g., geothermal plant)
Heat Q_2 added to warm reservoir at T_2	4 $\eta\ (COP) = Q_2/|\Delta H|$ $\varepsilon = \eta\ (1-T_0/T_2)$ (e.g., electrically driven heat pump)	5 $\eta\ (COP) = Q_2/|\Delta H|$ $\varepsilon = (1-T_0/T_2)(Q_2/B)$ (e.g., engine driven heat pump)	6 $\eta\ (COP) = Q_2/Q_1$ $\varepsilon = \eta\ \dfrac{(1-T_0/T_2)}{(1-T_0/T_1)}$ (e.g., solar heater)
Heat Q_3 extracted from cool reservoir at T_3	7 $\eta\ (COP) = Q_3/W_{in}$ $\varepsilon = \eta\ [(T_0/T_3) - 1]$ (e.g., electric refrigerator)	8 $\eta\ (COP) = Q_3/|\Delta H|$ $\varepsilon = \dfrac{Q_3}{B}\left[\dfrac{T_0}{T_3} - 1\right]$ (e.g., gas powered air conditioner)	9 $\eta\ (COP) = Q_3/Q_1$ $\varepsilon = \eta\ \dfrac{(T_0/T_3)-1}{1-(T_0/T_1)}$ (e.g., absorption refrigerator)

a. Source: AIP, 1975. Used with permission.

While a limited number of components are addressed here, it is emphasized that virtually any process can be analyzed with the concept of availability. Several references treat the topics discussed here in more depth. A study of some of these other works could be quite valuable (Van Wylen and Sonntag, 1985; Ahern, 1980; Kestin, 1980; Kotas, 1980a,b; Bejan, 1982; Moran,1982; Michaelides, 1984; Kotas, 1985).

7.3.2 Turbine Process

Consider a general turbine process in which a fluid stream enters at some thermodynamic state and flow rate. Assume that the process is taking place at steady state, so there is an equal flow rate of fluid exiting the device at another thermodynamic state. Accompanying this flow of mass, there is net power produced and, in general, heat exchange with the surroundings at T_o. In real machines, there will also be several forms of friction as well as other irreversibilities. A schematic of this device is shown in Figure 7.3, where the various states and energy flow quantities are defined.

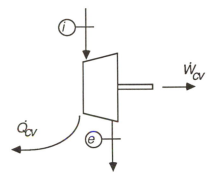

Figure 7.3 Schematic of a general turbine showing the various energy flow quantities.

Now apply Equation 7.3. The first term on the left-hand side represents the power output, which is assumed to be specified here. If this quantity is not given, it could be found using the First Law of Thermodynamics. The second term, the rate at which lost work is being produced, is not known initially; and this will have to be found from Equation 7.3 if a value is needed. Values for the terms represented by the first summation on the right-hand side can be calculated from the thermodynamic states of the flow streams entering and leaving and by using the known mass flow rates. Equation 7.2 can be applied. For most real systems, the final set of terms of the equation, those related to the rates of heat losses, is difficult to evaluate in precise detail. Sometimes an approximation to this last term will be made such that the rate of heat loss is taken as a single flow, with the corresponding temperature being an intermediate value between the entrance and the exit of the mass flowing through the turbine. The usual situation, however, is to assume that the surroundings are in contact with the turbine exterior and that the resulting heat transfer takes place in an externally reversible process to T_o.

It is easily shown that the First Law Efficiency is given by

$$\eta_{cv} = \dot{W}_{cv} / [\dot{m}(h_i - h_e)] \qquad (7.12)$$

When seeking an appropriate value for the Second Law Efficiency, there are several forms that can be used. Consider Equation 7.8. The numerator is the availability out, which is the actual power produced. Thus, Equation 7.8

149

would show the following to be the Second Law Efficiency:

$$\varepsilon_1 = \dot{W}_{cv} / [\dot{m} \, (\, b_i - b_e \,)] \qquad (7.13)$$

where

$$b = h - h_o - T_o(\, s - s_o) \qquad (7.14)$$

and any chemical reactions, kinetic energies, and potential energies are neglected. This is approximately the situation in square 1 of Table 7.2. It is stated there that $\eta = \varepsilon$. The input is not actually work here, but thermal energy transfer in an adiabatic process instead. A more accurate statement for this situation should be $\eta \approx \varepsilon$. It can be shown (de Nevers, 1981) that the two quantities are exactly the same for the special case of the average fluid temperature during the process being equal to the ambient temperature, T_o. Further information on expansion and compression processes has been given recently by Stecco and Manfrida (1986).

7.3.3 Heat Exchange Processes

A number of generic types of heat transfer processes occur in thermal systems. Within the general category of heat exchangers, there are cases where similar fluid flows occur on both sides; dissimilar flows occur on each side; and a change of phase occurs on one or both sides. There is, however, a multitude of situations in the design process where something other than a shell-and-tube heat exchanger may be needed. (See the discussion in Chapter 3 about the application of heat transfer devices.) Examples include furnaces and cooling towers as well as the various other configurations of direct-contact heat exchangers. Even simple heat transfer through a wall can be included in the general heat exchange processes. While some of the distinctions indicated here are artificial, numerous special situations could be considered.

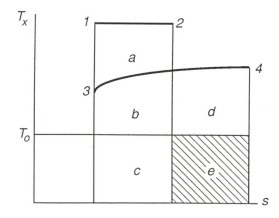

Figure 7.4 Comparison of an ideal and an actual heat transfer process. (After Ahern, 1980.)

7.3.3.a Some Basic Ideas

To introduce general ideas about the analysis of heat transfer processes, consider an ideal process and then consider its actual counterpart. In the ideal process, heat is transferred isothermally across an infinitesimally small temperature difference. This could be the process represented as line *1-2* in Figure 7.4. Since the temperature difference is infinitesimally small, the T_x represents the temperature of both the source and the sink. The area under line *1-2* is proportional to the total heat transfer. Hence,

$$Q_{1-2} = Area\ a + Area\ b + Area\ c \tag{7.15}$$

Of this total heat transfer, the availability can be interpreted as the maximum possible work that could be extracted from the process. The maximum possible work, in turn, is the amount that would result from a Carnot cycle operating between T_x and T_o. Consequently, for process *1-2*,

$$W_{max1-2} = Area\ a + Area\ b \tag{7.16}$$

This is also the availability, or

$$\Delta B_{1-2} = W_{max1-2} \tag{7.17}$$

Note that the amount of energy that would be rejected from the Carnot engine *(Area c)* is the unavailable energy.

Now consider the real (irreversible) counterpart of process *1-2*, namely, process *3-4*. Again it will be assumed that heat is transferred from an infinite source at T_x. This time the heat will be absorbed by the working substance at state *3*. For simplicity, the entropy of state *3* will be taken as the same value as state *1*, as shown in Figure 7.4. If it is required that the same amount of heat be transferred in process *3-4* as was transferred in process *1-2*, and assuming the process is internally (but not externally) reversible, then again the areas can be used to represent the heat transfer:

$$Area\ a + Area\ b + Area\ c = Area\ b + Area\ c + Area\ d + Area\ e \tag{7.18}$$

or

$$Q_{1-2} = Q_{3-4} \tag{7.18a}$$

The maximum work for the real process is

$$\Delta B_{3-4} = Area\ b + Area\ d \tag{7.19}$$

while the unavailable energy can be represented by *Area c + Area e*. The "lost work" can be represented as the difference in the unavailable energies between the ideal and actual processes. In Figure 7.4, this is denoted by the shaded *Area e*. Hence, the fact that the heat transfer takes place across a finite temperature difference results in an amount of lost work.

7.3.3.b Heat Exchanger Concepts

The discussion just given was meant to aid in developing physical insight into the various Second Law ideas previously noted. Now consider an example of how the availability quantities are calculated when considering a process block for a heat transfer device. In particular, focus on a conventional counterflow heat exchanger, shown schematically in Figure 7.5.

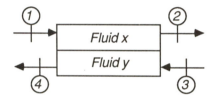

Figure 7.5 Heat exchanger notation used for illustrative example.

As is done with any type of process, application of Equation 7.3 will be made. For the heat exchanger considered here, no work is actually done. If it is further idealized that either there is no heat transfer with the surroundings or any heat transfer with the surroundings takes place in an externally reversible manner (the sink and the exterior of the heat exchanger are at the same temperature), then environmental heat transfer will not appear in Equation 7.3. Clearly, the two situations will be handled in different ways in the writing of the First Law. Assume that there is no external heat transfer. The First Law statement is written as follows for steady-state, steady-flow conditions:

$$\dot{m}_x (h_1 - h_2) + \dot{m}_y (h_3 - h_4) = 0 \qquad (7.20)$$

A Second Law statement can be expressed through Equation 7.3. According to the assumptions already made,

$$\dot{L W}_{cv} = \dot{m}_x (b_1 - b_2) + \dot{m}_y (b_3 - b_4) \qquad (7.21)$$

Continuing to assume that there are negligible kinetic and potential energy changes, as was implicitly done in Equation 7.20, then each of the availability terms is given by

$$b_j = h_j - h_{oj} + T_o (s_j - s_{oj})$$ (7.22)

Since there are no anticipated changes in chemical composition, then the chemical potential terms in Equation 7.2 are not incorporated here; and they would not have affected the results had they been considered. Combining the First and Second Law statements, the lost work becomes as shown in Equation 7.23. Note that the reference state enters only through the T_o quantity. This is a typical situation that occurs when lost work is calculated (de Nevers, 1981).

$$\dot{LW}_{cv} = T_o [\dot{m}_x (s_1 - s_2) + \dot{m}_y (s_3 - s_4)]$$ (7.23)

If thermodynamic property tables are accessible for fluids x and y, the values are easily found for the entropies. On the other hand, if thermodynamic properties are not available, but transport properties (like those found in the reference section of many heat transfer texts) are, then the entropy changes can be calculated for the special case of no pressure drop using specific heat values:

$$\Delta s_j \approx C_{pj} \ln (T_i / T_e)$$ (7.24)

There may be a need to determine a Second Law Efficiency for this heat exchanger. If so, then a form of Equation 7.8 can be used. Assuming that the desired result is the heating of fluid x with a hot fluid y, then

$$\varepsilon_1 = [\dot{m}_x (b_1 - b_2)] / [\dot{m}_y (b_3 - b_4)]$$ (7.25)

Nishio et al. (1980) use a way of representing heat transfer processes graphically that can be quite valuable. In this approach, the Carnot cycle factor $(1-T_o/T)$ is plotted against the heat transfer. Two examples of this are shown in Figure 7.6.

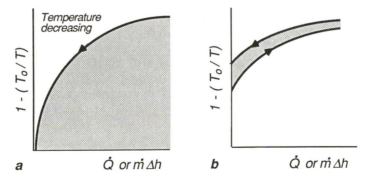

Figure 7.6 Plots of heat transfer processes. The dotted areas represent: **a**-the availability of the initial state, and **b**-the unavailable energy (Nishio et al.,1980).

In the two representations shown in Figure 7.6, consider the one on the left side (**a**). If the cooling takes place until the dead state is reached, the availability associated with the initial state is represented by the area under the curve. This can be seen by considering the definition of availability. If heat transfer takes place between two streams (such as in a heat exchanger), then the area between the two curves on the right (**b**) indicates the lost work or loss of available energy.

By using the same plot type as shown in Figure 7.6, the effect of pressure on the heat transfer to a fluid undergoing phase change can be visualized. An example is shown in Figure 7.7. The types of graphical representations shown in Figures 7.6 and 7.7 can be handy in visualizing the overall efficiency of heat transfer processes.

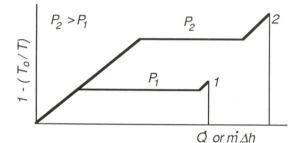

Figure 7.7 The effect of pressure can be shown on heat transfer processes in much the same manner as used in Figure 7.6. A phase change fluid is shown here. After Nishio et al. (1980).

Various studies reported in the literature have shown the utility of availability analysis in design considerations on heat exchangers. One of the earliest studies (McClintock, 1951) examined the trade-offs between decreasing the irreversibilities with increasing heat transfer areas. Other work on this same topic has been reported recently (Boyd et al., 1981; Bejan, 1982). Bejan has also investigated the question of how areas should be distributed when heating occurs over a large temperature range. He has shown that larger areas per unit temperature rise should be used as the absolute temperature of the fluid decreases if a goal is the minimization of irreversibilities. Chato and Damianides (1986) have dealt with the problem of optimizing locations of heat exchangers in systems.

A comprehensive summary has been given regarding the effect of a number of pertinent variables on the Second Law Efficiency for both parallel and counterflow exchangers (Golem and Brzustowski, 1977). It can be shown that an imbalance of the mass-flow-specific-heat products between the two streams can have a very large effect on the lost work generation (Bejan, 1982). Equations have been developed that show effects of various design parameters on irreversibility (London and Shah, 1983).

In a recent publication, availability analysis has been used to define optimal heat exchanger designs. In one, desirable designs for waste heat recovery/Rankine cycle heat exchangers were described (Tokuda and Osanai, 1984). In this work, a volume minimization was desired because the system was to be located on a ship. The authors showed that the best design was one that maximized the parameter $\varepsilon_{hex}/(UA/mC_p)$, where U is the overall heat transfer coefficient, A is the associated area, and the mass-flow-specific-heat product refers to the stream where this value is minimum. In another work (Tsujikawa et al., 1986), an optimal design of a gas turbine regenerator was accomplished using Second Law ideas.

7.3.3.c Forced-Flow Convection

Consider heat flow in forced convection, either an internal or an external configuration. The question might arise as to how this process might be made more efficient. Most engineers would ask additional questions about the meaning of the term "efficient," and with good reason. On the one hand, it is known that for a specific application, say a simple situation with fluid flow in a pipe, the heat transfer can usually be increased for a fixed piece of equipment and fixed temperature difference by increasing the effective velocity. On the other hand, it is also known that the irreversibility associated with fluid friction is a very strong function of the fluid velocity. This is shown in Equation 7.5. Since the "value" of heat flow may be considerably different than the "value" of fluid friction (heat vs. work), a consistent and realistic comparison of these two phenomena might be worthwhile.

A basis for comparison of heat flow irreversibilities and irreversibilities due to fluid friction is the concept of lost work introduced at the beginning of this chapter. Remember that a classical form of the Second Law for a process can be written as shown in Equation 7.26.

$$ds = (\delta q + \delta lw)/T \tag{7.26}$$

With this fact in mind, entropy production due to heat transfer and fluid friction (lost work) can be compared. For flow and heat transfer inside a pipe, it can be readily shown (Bejan, 1982) that the local entropy generation rate per unit length of pipe, S, is given by Equation 7.27.

$$S = \frac{q \, \Delta T}{T^2} + \frac{\dot{m}}{\rho T} \left[-\frac{dP}{dx} \right] \tag{7.27}$$

Here q is the local heat transfer rate per unit length of pipe, ΔT is the difference between the wall and the bulk fluid temperatures at the same location the q occurs, and T and ρ are taken as the bulk fluid values.

Using the definition of the Stanton number, St, and the friction factor, f, as defined in Equation 7.6, Equation 7.27 can be recast into Equation 7.28.

$$S = \frac{q^2}{4 \, T^2 \dot{m} c_p} \frac{D}{St} + \frac{2\dot{m}^3}{\rho^2 T} \frac{f}{DA^2} \tag{7.28}$$

In this equation, D is the hydraulic diameter of the pipe and A is the cross-sectional area of the pipe. Hence, for any pipe characteristics of the form

$$St = f_1(Re)$$

and

$$f = f_2(Re)$$

with Re being the Reynolds number, the entropy generation rate can be determined. This allows evaluations of the various trade-offs that might be available. Bejan (1982) develops this further to show an optimum Reynolds number, given by Equation 7.29 for the turbulent-flow case.

$$Re_{opt} = 2.023 \, [Pr]^{-0.071} \, [\rho m \dot{q} \, /\{ \mu^{2.5}(kT)^{0.5}\}]^{0.358} \qquad (7.29)$$

The rate of entropy generation increases quite sharply for Reynolds numbers that are either smaller or larger than the optimum value of Re.

This same approach can be used to analyze augmented surfaces (Oullette and Bejan, 1980; Bejan, 1982). Insights are possible about the effectiveness of these specially prepared surfaces when increased pumping power is traded off against increased heat transfer.

7.3.4 Psychrometric Processes (Wepfer et al., 1979b)

In the heating, ventilating, and air-conditioning field, the problem of dealing with air/water mixtures often arises. Consider a classical application of the air washer as represented by an adiabatic saturation process as shown in Figure 7.8.

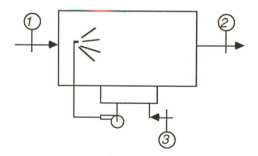

Figure 7.8 Schematic of an air washer. An air-water-vapor mixture flows from *1* to *2* and is sprayed with liquid water. Makeup liquid is introduced at *3*.

In the system shown in the figure, a flow of an air-water-vapor mixture enters at *1*, comes in contact with a liquid spray, and exits at *2*. Liquid is introduced to make up for the amount of water that is evaporated into the air flow through the device. Usually, the amount of work required

156

for the small pump can be neglected, and this will be done here. The pump work could be considered, if desired. It is assumed that there is negligible heat transfer to the surroundings. Further assume that the relative humidity, ω, the temperature, and the total pressure are all known at both *1* and *2*. If some unknowns exist, in general, they can be found from First Law considerations (Van Wylen and Sonntag, 1985), which will not be dealt with here. The temperature of the water introduced at *3* is also known.

The First Law balance on the air washer taken as the control volume and assuming steady-state, steady-flow conditions is

$$(h_{a1} + \omega_1 h_{v1}) + (\omega_2 - \omega_1)\, h_{f3} = (h_{a2} + \omega_2 h_{v2}) \tag{7.30}$$

where ω is the specific humidity, $\dot{m}_v / \dot{m}_a$, mass of water vapor over the mass of dry air in the same mixture. Casting the usual definition of availability into terms appropriate for the analysis of an air-water mixture,

$$b = h_a + \omega\, h_v - T_o\, [\, s_a(T,P_a) + \omega\, s_v(T,P_a)] - [h_a(T_o) + \omega\, h_v(T_o)]$$

$$+ T_o\, [\, s_a(T_o,P_{ao}) + \omega\, s_v(T_o,P_{vo})] \tag{7.31}$$

As is usually the case in systems operating near atmospheric conditions, the air is treated as an ideal gas. Entropy values for air are calculated relative to the reference state. Or

$$s_a(T_o,P_{ao}) = s_a(T_o,P_o) - R_a\, ln\,(P_{ao}/P_o) \tag{7.32}$$

R_a being the gas constant for air, which is the universal gas constant divided by the molecular weight of air. Note that, like the First Law form, Equation 7.31 is written on a unit-mass-of-dry-air basis.

The availability destruction can be written as the difference between that flowing into and that flowing out of the control volume. From Equation 7.3,

$$L\dot{W}_{cv} = \dot{B}_1 + \dot{B}_3 - \dot{B}_2 \tag{7.33}$$

while the efficiency of the device can be defined in terms of the increase of availability of the air stream divided by the availability of the water supply.

$$\varepsilon_1 = (\, \dot{B}_2 - \dot{B}_1)\, /\, \dot{B}_3 \tag{7.34}$$

By examining Equations 7.31 and 7.34, it is obvious that several terms will cancel in the numerator of the latter.

7.3.5 Combinations of Processes, Systems

Usually the designer deals with combinations of processes, or systems. While these can contain numerous components that may be best analyzed with flowsheeting (see the discussion of this concept in Chapter 8), a hypothetical system that contains three components will be dealt with here. This is shown in Figure 7.9.

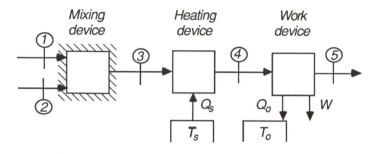

Figure 7.9 System to be analyzed for an overall amount of lost work.

The components considered include a *mixing device* where the working fluid at two distinct states and flow rates undergoes an irreversible adiabatic process. In the second component, a *heating device*, heat exchange occurs with a source at temperature T_s. Finally, there is a component where work is performed and the fluid undergoes a heat loss to the surroundings at T_o. This is denoted as the *work device*. Work can either be done <u>by</u> the work device or <u>on</u> the work device.

Assume that, if necessary, energy balances are performed on the system so that all thermodynamic properties are known at each numbered state. This will allow the analysis of each component much like has been shown in the earlier portions of this section. Alternatively, a control volume can be considered around the complete system, taking the work and heat interactions across the control volume boundaries. Each approach has its own advantages in separate instances. Sometimes it may be of value to examine where the irreversibilities occur within the system by examining each component, but other times an overall assessment may be desired. By analyzing a system both ways, the equivalence of the two methods will be demonstrated.

Considering a control volume around the complete system, the only energy quantities crossing the boundaries are the flows at *1, 2,* and *5* and the heat and work quantities. The heat transfer with T_o can be neglected in this analysis by assuming the process is externally reversible (i.e., the temperature outside the work device is T_o at the locations where the heat transfer takes place). Equation 7.3 becomes

$$\dot{W} + \dot{LW} = \dot{m}_1 b_1 + \dot{m}_2 b_2 - \dot{m}_5 b_5 + \dot{Q}_s [1 - T_o / T_s] \qquad (7.35)$$

All terms except the lost work would be known or could be determined from the First Law analysis. This evaluation may be sufficient if, for example, this system is to be compared to other systems that start with the same flow streams and produce a similar amount of work. Perhaps in one of these other systems, all the same general components are present; but the heat device is found in two components that heat both streams *1* and *2,* which are again followed by a mixing device, which is again followed by a work device.

Now consider an individual analysis of each component, similar to what was done in the earlier portions of this section. These balances are shown in a "spread" form here such that the various terms can be summed overall to give the lost work for the system.

Mixing device:

$$_{1,2}\dot{L}\dot{W}_3 = \dot{m}_1\, b_1 + \dot{m}_2\, b_2 - \dot{m}_3\, b_3 \tag{7.36}$$

Heating device:

$$_3\dot{L}\dot{W}_4 = \qquad\qquad + \dot{m}_3\, b_3 - \dot{m}_4\, b_4 \quad + \dot{Q}_s[1 - T_o/T_s] \tag{7.37}$$

Work device:

$$_4\dot{W}_5 + {}_4\dot{L}\dot{W}_5 = \qquad\qquad + \dot{m}_4\, b_4 - \dot{m}_5\, b_5 \tag{7.38}$$

Summing these equations together yields Equation 7.39.

Net:

$$_4\dot{W}_5 + \dot{L}\dot{W}_{cv} = \dot{m}_1\, b_1 + \dot{m}_2\, b_2 \qquad - \dot{m}_5\, b_5 + \dot{Q}_s[1 - T_o/T_s] \tag{7.39}$$

The equivalence between Equations 7.35 and 7.39 is readily apparent. The second method has the advantage of showing the location of the irreversibilities.

It is clear from this exercise that when the concept of lost work is incorporated, a Second Law balance between the flow availabilities, work terms, and modified heat transfer terms results. In this context, Second Law analyses can be made in a similar manner as First Law energy balances.

Second Law analyses can yield insights to the detailed performance of combined heat and work processes. The benefits of this approach have been demonstrated in a large number of situations. Examples include the analysis of combined gas turbine and steam power cycles (El-Masri, 1985) and many cogeneration situations that are discussed in the next section.

7.4 ENERGY AND AVAILABILITY ACCOUNTING

The cost of a plant, process, or component is of critical concern to the engineer. In the design of thermal systems, the problem of assessing costs becomes more involved compared to the purchase of general, manufactured commodities. As will be shown in this section, costs of the outputs of thermal systems can be based upon First and/or Second Law ideas. Often a First Law approach will yield satisfactory results, but at other times misleading conclusions might result. Hence, where there are trade-offs between heat and work to be evaluated, Second Law approaches should be used.

In many systems, the relative value of heat and work may need to be assessed. Prime examples of this are cogeneration systems, where both work and heat output have a commercial importance. There have been a number of empirical approaches to setting relative economic value on heat and work. (See, e.g., Comtois, 1978.)

For applications where the trade-offs between heat and work are desired, the investment in equipment can almost always be determined in a straightforward fashion. See Chapter 6. In a similarly direct manner, the fuel costs can be evaluated. Operation and maintenance costs can be estimated; or, if the plant is on line, these values can be calculated from actual expenditures.

Even with all of this data, the question may arise: what is the cost per unit of power produced compared to the cost per unit of heat produced? The answer to the question may not be obvious. While the Second Law in general deals with these kinds of relative assessments, several authors have dealt specifically with the economic evaluations. The analyses have been applied to a range of systems far broader than cogeneration concepts.

Some of the earliest studies of this type were performed in the 1960s and early 1970s (Tribus et al., 1966; El-Sayed and Evans, 1970). Later, Gaggioli and co-workers developed aspects of this technique for applications (Gaggioli, 1977; Reistad and Gaggioli, 1980; Gaggioli and Wepfer, 1980). Summaries have appeared (Moran, 1982, Kenney, 1984; Tsatsaronis and Winhold, 1985a; Kotas, 1985, 1986). Detailed analyses have been given for the evaluation of optimal allocation of steam and power in chemical plants (Nishio et al., 1982); feedwater heater replacement economics (Fehring and Gaggioli, 1977); proper sizing of steam piping and insulation based upon Second Law ideas (Wepfer et al., 1979a); comparing steam and electric drive option costs for boiler feedpumps (Gaggioli and Fehring, 1978); and the trade-offs in equipment performance versus price for power plants (London, 1982; Tsatsaronis and Winhold, 1985b). Related treatments of other applications have also appeared (Bloomster and Fassbender, 1980). A graphical approach for plant analysis is described by Zheng et al. (1986). While the view has been given that "exergy cannot be used for thermoeconomic analyses" (Szargut, 1980), this basis has yielded some insights not possible through other approaches.

To start the analysis, some authors have used what they call a "money balance" (Reistad and Gaggioli, 1980). The cost of a product that requires energy inputs consists of many factors as shown in Equation 7.40.

$$C_{prod} = \Sigma \, C_i \qquad\qquad (7.40)$$

where the C_{prod} is the total money expended in producing the desired output. The C_i factors represent capital equipment costs, fuel costs (these two are often dominant factors), operation and maintenance costs, and any other financial outlays that are required. It is convenient to define unit cost factors, denoted by $c_i \equiv C_i / P$, where P is the amount of product output. P may represent electricity production, process heating load, or any other thermally produced item. Assigning the notation F to the amount of fuel used to produce an item, Equation 7.40 becomes

$$c_{prod} = (\, c_{fuel} \, F + \ldots \,) / P \qquad\qquad (7.41)$$

The First Law Efficiency of a boiler process in a steam power plant is given by

$$\eta_{boiler} = P / F \qquad\qquad (7.42)$$

where, in this case, the P represents the power equivalence produced in terms of steam flow and F is the rate at which fuel is consumed.

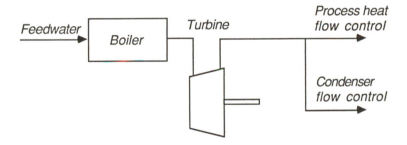

Figure 7.10 Simple schematic of a cogeneration system.

To develop the ideas further, consider a specific system, as given in Figure 7.10. Shown schematically is a boiler and a turbine. Depending upon the output needs, it is assumed that either shaft work or process steam or some combination can be derived from this arrangement. The turbine expansion is taken to various back pressures as a function of the process heat needed.

For example, assume a need for an increased amount of process heat and, within limits, the steam flow rate is constrained to be constant. A higher back pressure could be set on the turbine, and a higher specific enthalpy exhausting the turbine would then result due to the higher temperature and pressure there. A decrease in shaft power would be derived, and the high-energy exhaust steam could be used for the increased process heat requirement.

Conversely, if the shaft power is to be maximized, then the steam would be expanded to the maximum possible extent through the turbine. In this case, the steam would exhaust at the lowest possible temperature, shown here as the condenser. For the discussion here, denote the steam flow into the turbine as P_{hp}, the steam flow into the process stream as P_{ip}, and the steam flow into the condenser as P_{lp}. The unit cost of the high-pressure steam is set by the cost of the fuel and the cost of the boiler. Hence,

$$c_{hp} = \frac{c_f}{\eta_{boiler}} + \frac{C_{boiler}}{P_{hp}} \tag{7.43}$$

The total cost of the products is given by

$$C_{shaft\ work} + C_{lp} + C_{ip} = C_{expended} = c_{hp}\, P_{hp} + C_{turbine} \tag{7.44}$$

A similar approach is used to find the Second Law basis. A Second Law form that is similar to Equation 7.41 is given by Equation 7.45.

$$c_{prod} = (c_{fuel}\, A_f + \dots) / A_p \tag{7.45}$$

where c is the unit cost of the product based upon Second Law concepts. The Second Law form that is comparable to Equation 7.43 is given in Equation 7.46.

$$c_{prod} = \frac{c_{fuel}}{\varepsilon_{boiler}} + \frac{C_{boiler}}{A_{hp}} \tag{7.46}$$

Several observations can be made at this point. First, the unit cost of the product is considerably different than the unit cost of the fuel. Second, the unit cost of the product compared between the First Law analysis and the Second Law analysis can be similar or quite different, depending to a large extent upon the comparison of the numerical values of ε and η for the process. Examples of where these are quite different can be visualized by considering the difference between the availability and the work produced, and examples are cited in the literature (Reistad and Gaggioli, 1980). Finally, it should be noted that the unit cost of the products <u>might</u> be decreased by increasing the Second Law conversion efficiency, ε. However, because an increase of ε might require additional plant expenditures, such as more efficient pumps, larger heat exchangers, and so on, the actual unit cost of output might not decrease.

REFERENCES

AIP, 1975 "Second-Law Efficiency: The Role of the Second Law of Thermodynamics in Assessing the Efficiency of Energy Use," Chapter in EFFICIENT USE OF ENERGY (K. Ford et al., Eds.), American Institute of Physics, New York.

Ahern, J., 1980, THE EXERGY METHOD OF ENERGY ANALYSIS, Wiley-Interscience, New York.

Bejan, A., 1982, ENTROPY GENERATION THROUGH HEAT AND FLUID FLOW, Wiley, New York.

Bloomster, C., and L. Fassbender, 1980, "The Role of Second Law Analysis in Geothermal Economics," Energy, 5, pp. 839-851.

Boyd, J., V. Bluemel, T. Keil, G. Kucinkas, and S. Molinari, 1981, "The Second Law of Thermodynamics as a Criterion for Heat Exchanger Design," Energy, 6, pp. 603-609.

Bruges, E., 1959. AVAILABLE ENERGY AND THE SECOND LAW ANALYSIS, Academic, New York.

Chato, J., and C. Damianides, 1986, "Second-Law-Based Optimization of Heat Exchanger Networks Using Load Curves," International Journal of Heat and Mass Transfer, 29, pp. 1079-1086.

Comtois, W., 1978, "What is the True Cost of Electrical Power from a Cogeneration Plant?," Combustion, September, pp. 8-14.

Cozzi, C., 1975, "Thermodynamics and Energy Accountancy in Industrial Processes," Energy Sources, 2, pp. 165-178.

Darrieus, G., 1930, "The Rational Definition of Steam Turbine Efficiencies," Engineering, 130, 1930, p. 283.

de Nevers, N., 1981, "Two Fundamental Approaches to Second-Law Analysis," in FOUNDATIONS OF COMPUTER-AIDED CHEMICAL PROCESS DESIGN, VOLUME II (R. Mah and W. Seider, Eds.), AIChE, New York, pp. 501-536.

de Nevers, N., and J. Seader, 1980, "Lost Work: A Measure of Thermodynamic Efficiency," Energy, 5, pp. 757-769.

El-Masri, M., 1985, "On Thermodynamics of Gas-Turbine Cycles: Part 1— Second Law Analysis of Combined Cycles," Journal of Engineering for Gas Turbines and Power, 107, October, pp. 880-889.

El-Sayed, Y., and R. Evans, 1970, "Thermoeconomics and the Design of Heat Systems," Journal of Engineering for Power, January, pp. 27-35.

Fehring, T., and R. Gaggioli, 1977, "Economics of Feedwater Heater Replacement," Journal of Engineering for Power, July, pp. 482-489.

Gaggioli, R., 1977, "Proper Evaluation and Pricing of Energy," in ENERGY USE MANAGEMENT: PROCEEDINGS OF THE INTERNATIONAL CONFERENCE, VOLUME II (R. Fazzolare and C. Smith, Eds.), Pergamon, London, pp. 31-43.

Gaggioli, R. (Ed.), 1980, THERMODYNAMICS: SECOND LAW ANALYSIS, ACS Series 122, Washington, DC.

Gaggioli, R., and T. Fehring, 1978, "Economic Analysis of Boiler Feed Pump

Drive Alternatives," Combustion, January, pp. 35-39.

Gaggioli, R., and W. Wepfer, 1980, "Exergy Economics," "I. Cost Accounting Applications," pp. 823-832, "II. Benefit-Cost of Conservation," pp. 833-837, Energy, 5.

Golem, P., and T. Brzustowski, 1977, "Second-Law Analysis of Energy Processes, Part II: The Performance of Simple Heat Exchangers," Transactions of the CSME, 4, pp. 219-226.

Hevert, H., and S. Hevert, 1980, "Second Law Analysis: An Alternative Indicator of System Efficiency," Energy, 5, pp. 865-873.

Keenan, J., 1932, "A Steam Chart for Second Law Analysis," Transactions of the ASME, 54, p. 195.

Kenney, W., 1984, ENERGY CONSERVATION IN THE PROCESS INDUSTRIES, Academic, New York.

Kestin, J., 1980. "Available Work in Geothermal Energy," Chapter 3 in SOURCEBOOK ON THE PRODUCTION OF ELECTRICITY FROM GEOTHERMAL ENERGY (J. Kestin et al., Eds.), DOE/RA-28320-2, March. (Available from NTIS.)

Kotas, T., 1980a,b, "Part 1, Exergy Concepts of Thermal Plant," pp. 105-114, and "Part 2, Exergy Criteria of Performance for Thermal Plant," pp. 147-163, International Journal of Heat and Fluid Flow, 2.

Kotas, T., 1985, THE EXERGY METHOD OF THERMAL PLANT ANALYSIS, Butterworth, London.

Kotas, T., 1986, "Exergy Method of Thermal and Chemical Plant Analysis," Chemical Engineering Research and Design, May, pp. 212-229.

Kreith, F., and J. Kreider, 1978, PRINCIPLES OF SOLAR ENGINEERING, McGraw-Hill, New York.

London, A., 1982, "Economics and the Second Law: An Engineering View and Methodology," International Journal of Heat and Mass Transfer, 25, pp. 743-751.

London, A., 1985, "Irreversibility Analysis--A Methodology for Both the Designer and Generalist," ASME/AIChE Max Jakob Award Lecture.

London, A., and R. Shah, 1983, "Cost of Irreversibilities in Heat Exchanger Design," Heat Transfer Engineering, 4, no. 2, pp. 59-73.

Marin, J., and J. Turegano, 1986, "Contribution to the Calculation of Chemical Exergy in Industrial Processes (Electrolyte Solutions)," Energy, 11, pp. 231-236.

McClintock, F., 1951, "The Design of Heat Exchangers for Minimum Irreversibility," ASME Paper 51-A-108.

Michaelides, E., 1984, "The Second Law of Thermodynamics as Applied to Energy Conversion Processes," International Journal of Energy Research, 8, pp. 241-246.

Moran, M., 1982, AVAILABILITY ANALYSIS: A GUIDE TO EFFICIENT ENERGY USE," Prentice Hall, Englewood Cliffs, NJ.

Nishio, M., K. Shiroko, and T. Umeda, 1980, "A Thermodyamic Approach to Steam-Power System Design," I&EC Process Design & Development, 19, pp. 306-312.

Nishio, M., K. Shiroko, and T. Umeda, 1982, "Optimal Use of Steam and Power in Chemical Plants," I&EC Process Design & Development, 21, pp. 640-646.

Oullette, W., and A. Bejan, 1980, "Conservation of Available Work (Exergy) by Using Promoters of Swirl Flow in Forced Convection Heat Transfer," Energy, 5, pp. 587-596.

Petela, R., 1964, "Exergy of Heat Radiation," Journal of Heat Transfer, May, pp. 187-192.

Reistad, G., and R. Gaggioli, 1980, "Available-Energy Costing," in THERMODYNAMICS: SECOND LAW ANALYSIS, ACS Symposium Series, No. 122 (R. Gaggioli, Ed.), pp. 144-159.

Rodriguez, L., 1980, "Calculation of Available-Energy Quantities," in THERMODYNAMICS: SECOND LAW ANALYSIS, ACS Symposium Series, No. 122 (R. Gaggioli, Ed.), pp. 39-59.

Shieh, J., and L. Fan, 1982, "Estimation of Energy (Enthalpy) and Exergy (Availability) Contents in Structurally Complicated Materials," Energy Sources, 6, pp. 1-46.

Siemons, R., 1986, "Interpretation of the Exergy Equation for Steady-Flow Processes," Energy, 11, pp. 237-244.

Stecco, S., and G. Manfrida, 1986, "Exergy Analysis of Compression and Expansion Processes, Energy, 11, pp. 573-577.

Szargut, J., 1980, "International Progress in Second Law Analysis," Energy, 5, pp. 709-718.

Tokuda, S., and T. Osanai, 1984, "Exergy Recovery with an Exhaust Gas Economizer System," Bulletin of the JSME, 27, September, pp. 1944-1950.

Tribus, M., R. Evans, and G. Crellin, 1966, "Thermoeconomic Considerations of Sea Water Demineralization," Chapter 2 in PRINCIPLES OF DESALINATION (K. Spiegler, Ed.), Academic, New York.

Tsatsaronis, G., and M. Winhold, 1985a, "Exergoeconomic Analysis and Evaluation of Energy-Conversion Plants--I. A New General Methodology," Energy, 10, pp. 69-80.

Tsatsaronis, G., and M. Winhold, 1985b, "Exergoeconomic Analysis and Evaluation of Energy-Conversion Plants--II. Analysis of a Coal-Fired Steam Power Plant," Energy, 10, pp. 81-94.

Tsujikawa, Y., et al., 1986, "Thermodynamic Optimization Method of Regenerator of Gas Turbine with Entropy Generation," Heat Recovery Systems, 6, pp. 245-253.

Van Wylen, G., and R. Sonntag, 1985, FUNDAMENTALS OF CLASSICAL THERMODYNAMICS (THIRD EDITION), SI VERSION, Wiley, New York.

Wepfer, W., R. Gaggioli, and E. Obert, 1979a, "Economic Sizing of Steam Piping and Insulation," Journal of Engineering for Industry, November, pp. 427-433.

Wepfer, W., R. Gaggioli, and E. Obert, 1979b, "Proper Evaluation of Available Energy for HVAC," ASHRAE Trans., 85, Pt. 1, pp. 214-230.

Zheng, D., Y. Uchiyama, and M. Ishida, 1986, "Energy-Utilization Diagrams for Two Types of LNG Power-Generation Systems," Energy, 11, pp. 631-639.

PROBLEMS

If a problem is ill- or incompletely defined, make and state any necessary assumption(s) needed to find the information requested.

7.1 Determine the lost work that results for a simple counterflow heat exchanger process. Start by using Equation 7.24 for each of the streams. Neglect any pressure loss. Include a pressure drop of ΔP_j for each of the streams, which start at a pressure P_j. In both cases, take the mass-flow-specific-heat products to be finite and constant throughout the exchanger.

7.2 Consider a simple counterflow heat exchanger with mass-flow-specific-heat products $m_j C_{pj}$ for each of the streams. Neglecting any pressure drop effects, determine the lost work that occurs as a function of the ratio of the two mass-flow-specific-heat products.

7.3 400 lbm/hr of hot air are available at 500°F. It is desired to use this air to make steam at 100 psia, starting with water at that pressure and a temperature of 80°F. Assume that the following overall heat transfer coefficients are possible: preheating, 100 Btu/hr ft² °F; boiling, 200 Btu/hr ft² °F; and superheating, 50 Btu/hr ft² °F. Specify the heat exchanger areas required for each of the three regions to produce the maximum amount of steam. What is the lost work associated with this process? Describe some ways that the lost work could be reduced.

7.4 An open feedwater heater is to be installed in a steam power plant. Feedwater at 1 atm pressure, 100°F, and 500,000 lbm/hr enters through one port. Through another port, dry saturated steam at 1 atm pressure is available at any flow rate desired. If the purpose of this direct-contact heater is to raise the feedwater to the highest possible temperature before the mixture of the feedwater and steam exits at 1 atm pressure, determine the flow of steam required. Calculate both a First Law Efficiency and a Second Law Efficiency for this device.

7.5 Compare the higher heating value with the availability of three gaseous fuels.

7.6 Calculate the First Law and Second Law efficiencies for vapor compression air conditioning and compare these values to their counterparts for an evaporative cooling application. The ambient conditions are dry-bulb temperature, 35°C, and relative humidity 15%. In the cooled space the desired temperature is 23°C. You may assume that the vapor compression machine operates at 60% of its Carnot performance and that the evaporative cooler can cool to 3 °C above the wet-bulb temperature. Neglect the power required to circulate the air in both cases.

7.7 Using appropriate relationships for the Stanton number and the friction factor in Equation 7.28, determine the entropy generation rate per unit length of pipe for turbulent flow and negligible pipe wall roughness. Consider the fluid to be air at 1 atm and 100°C. Show the variation with pipe diameter and air mass flow rate in an appropriate plot. Investigate two values of the heat-transfer-coefficient-ΔT product, where the temperature difference is that between the bulk fluid and the tube wall.

7.8 A simple organic Rankine cycle is to be designed using R-113 as the working fluid. In the design analysis, attention is focused on the turbine. R-113 enters the turbine at 200°C and 3 atm pressure. If the condenser pressure is 0.6 atm, and the turbine has an adiabatic (First Law) efficiency of 0.82, find the Second Law efficiency of the turbine. Also determine the lost work/kg for this process. In a similar situation, everything remained the same, except that 10% of the enthalpy change through the device is lost to the surroundings (T_o=30°C) as heat transfer, find the Second Law efficiency of the turbine and the lost work/ kg.

7.9 Model the availability equations for both a gas-fired water heater and a similar water heater that utilizes electricity. Assume a 10% rate of standby heat loss compared to the heating load. Calculate the Second Law efficiencies for the two cases when propane is used as the gaseous fuel. For the gas-fired unit, assume 80% of the chemical energy in the fuel is transferred to the water.

CHAPTER 8
SYSTEM FLOWSHEETING

8.1 INTRODUCTION

When modeling any thermal system involving a number of process components, a flowsheeting approach will almost always be used. Having said that, the term *flowsheeting* should be defined. Flowsheeting will be taken here as a block diagram representation of a physical system. In this form, components (or process elements) are connected together at points where mass flows occur between those components. There could be other kinds of connections, such as those where energy or availability flows occur, but only mass flow representations will normally be denoted here. However, keep in mind that mass flows also represent enthalpy and availability flows implicitly. If a need exists to indicate only energy flows, that will be satisfied by rendering a separate diagram or by indicating additional aspects on the mass flow diagrams.

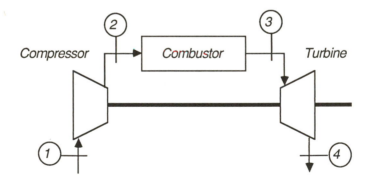

Figure 8.1 Schematic of a Brayton cycle. This is an example where the system can often be analyzed in a sequential manner where calculations of state variables begin at state *1* and proceed through state *4*.

The task of simply drawing the flowsheet can be quite challenging in certain situations, and the accomplishment of this may be valuable in its own right. This is a special area of research emphasis. See, for example, Curtis et al. (1981a,b). Normally, though, this is only the first step in a series leading to an analysis of a system. At one level, the analysis of systems employing flowsheets is very familiar to every student of engineering thermodynamics. Consider a simple Brayton engine cycle in a block diagram representation. See Figure 8.1. This is as it is shown typically in beginning texts. The student might be asked to determine the various state points and energy flows from the given information of the problem, say incoming temperature and pressure, compressor and turbine efficiencies, and pressure ratio. Normally, the solution process is to go through the calculations sequentially, probably starting at *1* and working around the figure.

While this is an example of system flowsheeting, it is a special case. This is true because of two aspects. For one, the system is quite simple compared to many because there are only three components. Another facet that makes this different from many of the problems encountered is that the calculations can proceed *sequentially*. Often, a series of conditions can only be determined *simultaneously*. A very simple example of this is shown in Figure 8.2. Here a system is taken to be a pump that forces fluid through some hydraulic resistance. Both the pump head curve and the pressure drop characteristics of the load are assumed to be known. One must solve for the resulting flow rate from the two relationships simultaneously.

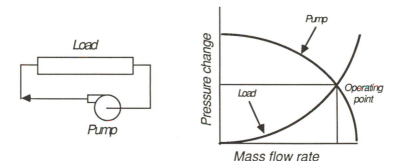

Figure 8.2 An example of a system that cannot be solved directly in a sequential manner. Here the operating point must be determined by solving simultaneously between the equations for the pump and the load.

The development of procedures for flowsheeting is receiving a great deal of attention in a number of engineering fields. For some time now, workers in electrical engineering have been developing systems theory suitable for the analysis of circuits and related problems. See, for example, Roe (1966). Some researchers have specifically applied the electrical circuit notation to the solution of problems involving thermal systems (Chinneck and Chandrashekar, 1984). Another major thrust, which is related to the circuits work but is more closely aligned to the topic of this text, is found in the chemical engineering field. Several excellent reviews have appeared in the literature (Motard et al., 1975; Westerberg et al., 1979; Rosen, 1980; Evans, 1980).

In what follows, an introduction to this topic is given. Because the chemical engineering field involves, or is closely related to, the design of thermal systems, the treatment here will be given more from that point of view. Enough information is furnished to allow simple modeling procedures to be set up. Readers who are interested in expanding their skills beyond what is given here should consult the many references noted at the end of this chapter.

8.2 THE NEED FOR A CONSISTENT NOTATION

When flowsheeting is first considered, it becomes apparent quickly that one of the most important needs is a versatile system for notation. It is possible that a given plant design may have hundreds of components, virtually every one connected to at least two others, with numerous cross-connections, flow splits, and mixing elements, and with heat flow from some components to others, and so on.

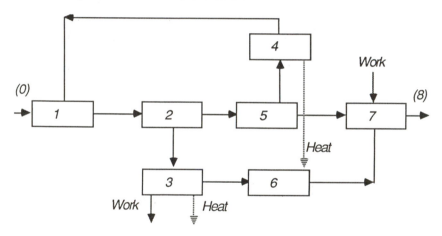

Figure 8.3 Diagram of a hypothetical system used to describe a possible notational method.

Consider the hypothetical example shown in Figure 8.3. Here a notation of numbering each component is used. The order chosen to assign numbers has no particular importance. (In very large systems, it may be desirable to use a more involved set of symbols that can be keyed to the coordinates on a detailed layout of the plant.) While this is by no means the only way to notate a diagram, nor is it probably even the best, it will illustrate some considerations that might be given in developing the more important features.

Figure 8.3 shows mass flow streams and, implicitly, the energy and availability carried with the mass, with the heat and work interactions shown secondarily. It is normally the case that the diagram will be based upon the various process elements and where the mass flows connect. Heat and work interactions are shown as distinct types of lines. In the convention used here, the crossing of heat, work, or mass lines with other mass flow lines is of no significance.

Situations arise from time to time where flowsheets are of value to represent solely energy or availability flows or the lines that denote flow of a particular chemical specie. Examples of these circumstances are not dealt with here, but they are not fundamentally different than those that are discussed. Keep in mind, though, that a diagram that shows mass flows will also be the same as a diagram of energy and availability flows that are carried with the mass.

The next item to emphasize is that mass flow lines do not connect directly with one another. Clearly, there might be situations where two streams could mix or split, for example, at a pipe tee. If this happens, a component is shown. Blocks *1, 2,* and *5* in Figure 8.3 are examples of these types of components. Also, whenever there is a change of state (temperature, pressure, etc.), a component box is shown. A box can be included for a piping run to represent real effects like pressure drop and/or heat transfer.

As the diagram becomes filled with several components, it happens quite frequently that mass flow lines must cross. Again it is emphasized that the mass flow streams do not connect (except at a component box), but they can cross.

Finally, it is desirable to have shorthand notation to denote the specific chart location of various simulation variables: mass flow rate, pressure, temperature, enthalpy, entropy, availability, species, and so on. In the simple systems dealt with in introductory thermodynamics texts, the state point is normally represented by a single number between components. The situation shown in Figure 8.1 is an example of this. While this approach is very straightforward and simple to use (and will be used for illustrative purposes here from time to time), when the system becomes more complex, several drawbacks become apparent. One of these is that components need to be denoted. While the components could be named by combining the numbers of the streams that enter and leave each component, this can become quite cumbersome. Hence, numbering of streams is not desirable. The approach noted in conjunction with Figure 8.3, where components are numbered, is more workable.

<u>Two very critical general sets of items regarding the processes are not shown on the diagram but should be noted</u>. The first of these is a mathematical description of the physical operation of the various components. For example, if the device is a heat exchanger, an Effectiveness-NTU formulation might be used to describe the unit's operation. Some units' descriptions can be extremely complex.

Second, it should be kept in mind that in almost all components there are physical parameters that must be set. In a flow splitter, the fraction of the flow that goes in each direction must be specified. Even in a simple pipe flow pressure drop calculation, the roughness and diameter of the piping must be given. Depending upon convention, the pipe roughness and diameter might be considered part of the physical model of the device, or they might be considered as independently set parameters related to the model. More insight about the representation of these aspects is given in the next section.

8.3 MATHEMATICAL DESCRIPTION

The introduction of a simple mathematical description of the flowsheeting process will be of value at this point. A more comprehensive discussion of many of these topics has been given by Westerberg et al. (1979).

Each of the components will have a behavior that can be described mathematically. The functional form is denoted here by F_i for the ith component. One example of this is the mathematical formulation of the thermodynamic model of a steam turbine in a Rankine cycle. This model should relate to the energy and entropy balances as well as the mass balance on the device.

A critical set of inputs to the formulation is the *flow information*. This includes the mass flows as well as the energy carried with those flows. $x_{i,j}$ will be used to denote these variables. Hence, x could mean (mass flow rate), (mass flow rate) · (enthalpy), (mass flow rate) · (entropy), a similar form involving availability, or a combination of these items, depending upon the type of model being developed.

In addition, there generally could be several *design or performance parameters* that are set for a specific component. Some relative efficiency, flow split fraction, approach temperature difference, or other appropriate variable would be examples. This description will be noted by y_i for the ith component.

Thus, a general set of thermodynamic process components that make up a flowsheet can be represented mathematically as

$$F_i (x_{0,1}, x_{1,2}, \ldots, x_{i,j}, \ldots, x_{m,n}, y_i) = 0 \qquad i = 1, n \qquad (8.1)$$

There is one of these functions for each of the n components on the flowsheet as denoted by the index i.

The exterior components, like *0* and *8* in Figure 8.3, might yield the extreme values of the parameter i, or they might be simply described components like the atmosphere or perhaps a source of fuel and not given a specific component function. Also, in certain situations, the variable y_i could denote a very complex functional form of its own. Finally, the notation is used here such that $x_{i,j} = - x_{j,i}$. Others have used a distinct variable for outflows compared to that used for inflows. In some instances, this may be desirable, but it tends to yield an additional set of equations where the outflow from one component is set equal to the inflow to the next component on the same flow path.

8.4 TWO APPROACHES

Two of the more commonly used approaches to the solution of flowsheeting problems will be described here. While almost any thermal system problem could be solved with either of these techniques, it will become apparent that certain kinds of problems and certain kinds of computer capabilities might swing the pendulum in favor of one or the other of these approaches. Outlined below are the trade-offs to be considered when contemplating a flowsheeting task.

8.4.1 The Sequential-Modular Approach

The *sequential-modular* technique was an outgrowth of the simplest calculations that were performed before the major use of computers. As the name implies, each process, or group of processes that could be combined into a module, is analyzed sequentially. Critical to the technique is that the outputs of each module (and the system as a whole) can be calculated from the inputs to the module (and the system). Hence, Equation 8.1 is solved for each component output in terms of the input(s). Referring to both Figure 8.3 and Equation 8.1, for example, the following could be written for component *1* :

$$x_{2,1} = f_1(x_{4,1}, x_{0,1}, y_1) \tag{8.2}$$

Here f_1 denotes the rearranged form of F_1 when Equation 8.1 becomes Equation 8.2.

(Note component *2* in Figure 8.3. With two outputs, it is necessary that each should be determined from the single input and functional specification.)

The characteristics of the sequential-modular approach that require the outputs be solved from the component inputs are both good news and bad news. On the good side, the calculation approach is often very straightforward and intuitive. It is used in virtually all beginning thermodynamics courses. The system shown in Figure 8.1 usually can be analyzed from given data, calculating from the input point to the output.

There are also some negative aspects to the sequential-modular approach. The approach does not allow specifications on the output to be made easily (sometimes referred to as *design* specifications). (See more discussion on this below.) Also, when there are recycle streams, like *4,1* in Figure 8.3, the solution technique can become quite complicated.

To illustrate this with an example, consider a fluid of interest (flow rate and inlet temperature specified) that flows through a bank of heat exchangers in series. At each heat exchanger, heat is removed from the primary fluid, and a pressure drop could occur. In this regard, consider Figure 8.4. If the problem is simply to analyze the output conditions of the primary fluid with each of the component heat exchanger's secondary fluids specifications fully given, then the answer follows directly from the use of the sequential-modular approach. On the other hand, if it is desired to determine the flows of secondary fluids (which have their temperatures specified) in each of the component heat exchangers such that the resulting temperature of the primary fluid exiting from the total bank is some specified value, the sequential-modular approach is less easily utilized. In this case, some estimates of the intermediate flows must be made, the end result checked, and an iteration performed if the answer is not the desired one.

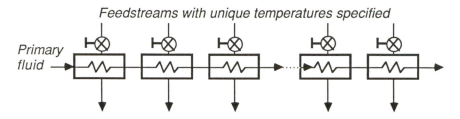

Feedstreams with unique temperatures specified

Primary fluid

Figure 8.4 Network of heat exchangers that cools the primary fluid shown flowing horizontally.

The estimation of flows, or other conditions, is called *tearing;* and this is a frequently used technique in this approach. Be concerned in the application of the sequential-modular technique to use tearing in an optimal manner. Needs for this arise from concerns for the ability to solve the problem in general, but the tearing also can affect the general efficiency of the solution. A good review of this has been given by Rosen (1980).

Another point with significant implications for this technique is the *degrees of freedom* in a given stream. This term means the number of independent parameters must be specified to define the flow stream fully. From thermodynamics, it is known that the degrees of freedom for a single component stream is equal to two intensive thermodynamic properties (often these will be pressure and temperature) and the flow rate. This statement assumes chemical and phase equilibria are present. When several components are present, the situation can become involved. The reader is directed to one of the references where this is described in detail to see what complications can arise. See, for example, Westerberg et al. (1979).

Now consider again the system shown in Figure 8.3. Examine the implications of each component denoted there in terms of a sequential-solution technique.

An important point needs to be noted again and emphasized here. If there are specifications given about the output stream(s) in a sequential calculation, this can cause a great deal of difficulty in the calculational procedure. This might show up in the example depicted in Figure 8.3 and Table 8.1 in the following manner. Suppose that details about the output(s) *7,8* were specified. In general, this approach would require assumption(s) to be made about the value(s) for the input(s) *0,1*. Then the whole scheme outlined in the table would be followed. A check of the *7,8* values thus calculated would then be compared to value(s) that were originally specified. If the value(s) did not check, a complete recycle through the calculations would be necessary. The situations where output variables are specified are sometimes denoted as the use of *design specifications.*

Of course, there could be special instances where complete recycles are not necessary. For example, consider the situation in the above system where there is no accumulation of mass and the specification on *7,8* might be a given mass flow rate. This allows the use of a mass flow rate of unity at *0,1*; the calculational procedure outlined in Table 8.1 to be followed; and the input and all other flows to be scaled as appropriate to yield the desired output. If there had been a mass accumulation within the system, or if the output specification had been on some other variable than mass flow at *7,8*, then the more general recycle situation arises.

Table 8.1
Sequential Calculation Procedure for the System Shown in Figure 8.3

State Number	Input	Output	Can Solve Directly?	Comments
1	0,1 & 4,1	1,2	No	Assume value(s) for 4,1.
2	1,2	2,5 & 2,3	Yes	
5	2,5	5,4 & 5,7	Yes	
4	5,4	4,1 & Q	Yes?	Yes, if $Q_{4,6}$ is given explicitly.

If calculated 4,1 value(s) found in the calculations here do not compare to the assumed value(s) in component 1, it must recycle through the above calculations with the new value(s) for 4,1.

State Number	Input	Output	Can Solve Directly?	Comments
3	2,3	3,6 & Q & W	Yes	
6	3,6 & Q	6,7	Yes?	Yes, if $Q_{4,6}$ is given explicitly.

If $Q_{4,6}$ is not given explicitly, but perhaps calculated from the difference in temperatures of state points 4 and 6, then a value for $Q_{4,6}$ would have been assumed in state 4 and the value would have been checked here. If the check does not yield the same value, all of the steps above would be repeated.

State Number	Input	Output	Can Solve Directly?	Comments
7	5,7 & 6,7 & W	7,8	Yes	

Overall mass and energy balances can be formed at this point to see if they check. Assuming no mass accumulators, at steady state the mass flow rate at 0 would have to be the same as at 8. The difference in energy flows at these two points must equal the net of the heat and work flows, again assuming no accumulation. If balances do not result, recycle through the complete calculational scheme.

A general approach to design specifications is to handle the overall set of processes affecting the specified output with a root-finding procedure. In this way, the overall set of elements is treated like a control block with an adjustable parameter that influences the output from the combination of blocks.

Typically, each component shown in the flowsheet will be represented as a subroutine in the overall calculational scheme. Thus, like components can use the same basic description of the physics of the process. Only the specification(s) of the parameter(s) and the input(s) for the individual components are needed for each set of calculations on each component to find the output value(s). These points are illustrated in Figure 8.5 to emphasize the ideas with the method of making the calculations sequentially. Examples of the application of this pattern of notation to sample process components are given in Section 8.5.

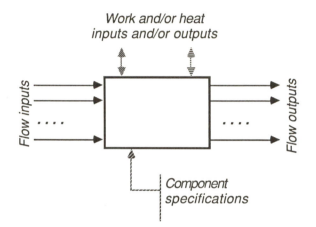

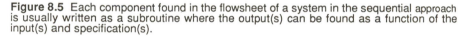

Figure 8.5 Each component found in the flowsheet of a system in the sequential approach is usually written as a subroutine where the output(s) can be found as a function of the input(s) and specification(s).

8.4.2 The Simultaneous-Solution Approach

The *simultaneous-solution* approach is the simplest to understand in mathematical terms, but it is often the most difficult to apply to actual systems. Here the various functional forms and unknowns are developed for a given set of processes. Generally there will be m equations and m unknowns. (Note that m will always be greater than the number of components in the system and could be <u>considerably</u> greater.) These equations will include not only the functional forms shown in Equation 8.1, but they will also include the equations of state and any other relationships necessary to represent the system fully in a mathematical sense. A solution approach that solves simultaneous nonlinear (most of the equations will be highly nonlinear) equations is then instituted. All of the design variables and size parameters are found in this manner.

The approach sounds quite easy. Sometimes it is. Often, though, the solution technique is very complicated. It is possible to have a system of thousands of nonlinear equations in the same number of unknowns. Another problem is that the system may be so configured that no solution is possible. An error, either typographical or logical, may yield a negative absolute temperature or some other preposterous situation.

It is necessary in the solution of systems of nonlinear equations to have starting values to find the solution. Determining starting values can also pose complications.

One of the major benefits of the use of a simultaneous solution to the complete system of equations is the ability to handle specifications on the output or any intermediate stream(s). This aspect generally presents no complication in the solution, unless, as noted above, the resulting solution is actually impossible.

To see how the simultaneous-solution technique works, consider again the system shown in Figures 8.1 and 8.2 and set up the governing equations in a form that could be solved by a simultaneous-solution technique. The Brayton cycle flowsheet is shown with a new numbering scheme in Figure 8.6, while the flow system from Figure 8.2 is shown in Figure 8.7.

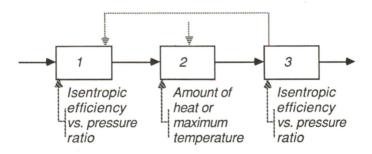

Figure 8.6 Another representation, here more typical of flowsheeting approaches, of the Brayton cycle system shown in Figure 8.1. For this version a new numbering system is used, with *1* representing the compressor process, *2* denoting the burner, and *3* indicating the turbine.

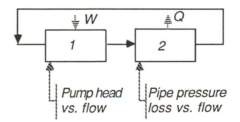

Figure 8.7 Another representation of the pump/piping circuit shown in Figure 8.2.

The Brayton cycle representation for simultaneous solution is not too different than it was before, except for the difference in the way the components are denoted. Now the numbers are on the components, as shown

177

in Figure 8.6, not between them, as shown in Figure 8.1. Assume that the analysis is to be performed for an ideal cycle with the use of constant specific heats. Assume further that the following data are given for the analysis:

$$P_{0,1}/P_{1,2} = P_{3,4}/P_{2,3} = \text{pressure ratio (given)}$$
$$P_{0,1} = P_{atm} = \text{atmospheric pressure (given)}$$
$$T_{0,1} = T_{atm} = \text{atmospheric temperature (given)}$$
$$c_p = \text{specific heat at constant pressure (given)}$$
$$c_v = \text{specific heat at constant volume (given)}$$

It is desired to find the maximum temperature in the cycle (i.e., $T_{2,3}$) such that the cycle thermal efficiency is some given value η. While it may be possible to combine all of the governing relationships in the problem algebraically to solve for the unknown temperature, a more general approach will be taken here.

The problem is solved by first writing the governing equations, some of which may be results of the First or Second Laws of thermodynamics. Unknown variables are denoted below by boldface characters.

$$k = c_p/c_v$$

Component 1 $\quad \boldsymbol{T_{1,2}}/T_{0,1} = (P_{1,2}/P_{0,1})^{[k-1]/k}$
$$\boldsymbol{w_c} = c_p(T_{0,1} - \boldsymbol{T_{1,2}})$$

Component 2 $\quad \boldsymbol{q_h} = c_p(\boldsymbol{T_{2,3}} - \boldsymbol{T_{1,2}})$

Component 3 $\quad \boldsymbol{T_{2,3}}/\boldsymbol{T_{3,4}} = (P_{2,3}/P_{3,4})^{[k-1]/k}$
$$\boldsymbol{w_t} = c_p(\boldsymbol{T_{2,3}} - \boldsymbol{T_{3,4}})$$
$$\eta = 1 - (\boldsymbol{w_t} + \boldsymbol{w_c})/\boldsymbol{q_h}$$

The known values are now substituted into the governing equations, and the unknowns (shown above as boldface characters) are determined from the given information. The unknowns are $\boldsymbol{T_{1,2}}, \boldsymbol{T_{2,3}}, \boldsymbol{T_{3,4}}, \boldsymbol{w_c}, \boldsymbol{w_t}$, and $\boldsymbol{q_h}$. This reduces to the equivalent of a linear system of six equations and six unknowns. A solution should follow in a straightforward manner.

In a similar manner, the pump/piping system shown in Figures 8.2 and 8.7 can be easily set up for solution via a simultaneous-equations approach. If it is assumed that the system is handling an incompressible substance at essentially isothermal conditions, then an equation of state is not needed. Two nonlinear equations result.

$$F_1(\boldsymbol{m}, \Delta\boldsymbol{P}, \text{pump characteristics}) = 0$$
$$F_2(\boldsymbol{m}, \Delta\boldsymbol{P}, \text{piping characteristics}) = 0$$

Using a formal technique, the two equations give the $(m, \Delta p)$ solution.

8.4.3 Comparing the Sequential and Simultaneous Approaches

Numerous differences between the two techniques (sequential modular and simultaneous solution) have just been introduced. A summary examination of some of these points will be of value. It is important to understand the critical differences between the two methods so that you can be better prepared to choose between them for a given application.

First, it should be noted that the sequential-modular approach is made up of a number of components that can be defined, categorized, and analyzed, with each type of component as a separate subroutine. In general, this is a straightforward (some times more so than other times) exercise of writing the outputs of the block in terms of the inputs and component specifications. A physical correspondence exists between the items on the flowsheet and the subroutines in the executive program. Keep in mind that a set of routines can be developed, one for each of the generally encountered process elements. Pay careful attention to this point because it is one of the very desirable aspects of the sequential-modular approach.

On the other hand, the simultaneous-solution approach is not usually performed in this stepwise solution manner. Instead, the governing equations of the system are laid out, the knowns so indicated, and the unknowns found. It is very critical that a complete system of equations (neither fewer nor more equations than necessary) be found. This is sometimes a difficult task, with some of the equations appearing to be arbitrary or extraneous.

Third, the solution approaches can make either of the two techniques appear to be "best." System simulation can be considered quite simply to be the determination of the solution of n equations with n unknowns. However, because of the potential problems with finding the appropriate equations indicated in the paragraph above, this could be difficult. If the system makeup and given data are appropriate, all outputs might be found explicitly in terms of all of the inputs. This would make the sequential-modular approach easy to apply.

Fourth, decisions on which system to use when considering the sequential and simultaneous techniques might be tipped slightly toward the sequential-modular approach. The major difficulty with the sequential technique (namely the problems of handling recycle streams) has been given much research attention (Westerberg et al., 1979), and several successful approaches have been found. With the simultaneous-solution technique, the problem of completely defining an appropriate system of equations is often difficult to handle. Furthermore, the solution technique for simultaneous, nonlinear equations can sometimes be a difficult task.

Finally, it should be noted that a great deal of work has been done to try to combine the positive aspects of both the sequential and simultaneous approaches. It may be possible to use predefined process modules with an essentially simultaneous approach. Ways of accomplishing this are still being evaluated. See Rosen (1980) and Timar et al. (1984) for more information on these approaches.

8.4.4 Applications Experience

The use of flowsheeting in the chemical process industry has a long history. Related types of efforts have been present in the electrical engineering fields over a similar period.

As noted by Motard et al. (1975), the work on flowsheeting in the chemical process industries began in the mid-1950s with development of computer programs for individual unit operations. Successes in this development were limited, but they did exist. This work led to additional efforts in the early 1960s of system simulations, with some of the simulators whose development was initiated in that era still in use today. Characteristics of that work included: code development was done in a highly modular fashion and in a higher level language, often FORTRAN; correlations were used that were quite rigorous; and the form of the code was such that people who were not familiar with its development could be users. An example of this kind of development was reported by Rosen (1962): Outgrowths of that period work have included the widely used FLOWTRAN program (Rosen and Pauls, 1977) and the newer, and more powerful, ASPEN code (Evans, 1980; ASPEN, 1982).

Reviews of simulators have been given in numerous articles (e.g., Hlavacek, 1977; Rosen, 1980; O'Shima, 1980). The comprehensive book by Westerberg et al. (1979) also gives an excellent summary of work. This topic is one of research interest, so new articles are appearing frequently.

8.5 EXAMPLES OF MODULAR FORMS

For the final section of this chapter, some examples of the components that might make up a flowsheet are briefly described. Simply consider the overall flowsheet to represent any kind of thermal system of interest in the design/analysis stage. This may be a power plant, a waste heat reclaiming system, a steam-heated crystallizer, or any other kind of device envisioned to be made up of a series of thermodynamic processes. In this section, some of the more generally encountered components will be shown in a context of value in setting up a sequential-modular solution scheme as has been discussed above. Whenever balances are required, mass and/or energy formulations will be used. Keep in mind that an availability basis could also be used, as was discussed in an earlier chapter. For brevity, it will not be included here.

In each case given below, a simple physical representation of the component will be given to indicate the mass flows, heat and/or work flows, and information specifications. This will be followed by a simpler block diagram of the component. In all cases, the notation used here will be that mass flows will be shown on the right and left faces of the block. Heat and work interactions will be shown on the top of the block. Information specifications will be shown entering the bottom of the block. Similar diagrams can be easily devised for components not shown here.

8.5.1 Pump Representation

Pumps are assumed to process an entering mass flow stream at one pressure to the same flow at a higher pressure, and this is accomplished by the input of work. A number of specifications can be made to define the performance of the pump. Generally, however, this involves specifying either a pump efficiency or a pump flow versus head curve (this is assumed to include the special case of a given single value of pressure increase across the pump). These aspects are shown in Figure 8.8.

Since this is the introductory figure to a sequence of diagrams, some conventions that will be followed should be explained. First, there could be a variety of specifications given on the mass flow streams. Not only do the thermodynamic properties need to be specified in some way, but there could also be a number of chemical species flowing together, and these would either need to be specified or determined through the solution technique. Of course, the total or individual mass flow rates also would need to be known or found. Hence, little specific information about the mass flow streams is shown in Figure 8.8, but that is simply for the sake of brevity in the illustration. All of the following diagrams in this section will not address the specification of the mass streams either.

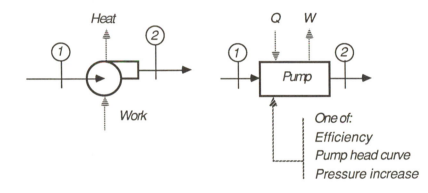

Figure 8.8 Block representation of a pump process.

Another aspect to note is that heat and work are shown flowing in opposite directions in the left-hand portion of the figure. Their directions are then reversed in the right-hand portion. This is done to emphasize the point that no matter which direction is present, the same convention is used throughout. From the overall system diagram, it may be obvious which direction the energy is flowing. Apparently, the left-hand diagram in Figure 8.8 reflects this. But because there are times when heat and work directions may not be apparent, the right-hand diagrams will always be shown with the heat flowing inward and the work flowing outward. It is recommended that this convention be used. Also, it should be noted that the flows indicated as work and heat probably will be work rate of flow (power) and rate of heat flow. This is simply a shortcut in notation.

8.5.2 Turbine Representation

Diagrams for a turbine, be it driven by steam or gas, are shown in Figure 8.9. Little comment is needed here, as the convention defined in Figure 8.8 is applicable.

A gas compressor follows an almost identical specification to that shown for the turbine, except for the obvious reversed directions in flows and work, and will not be shown here.

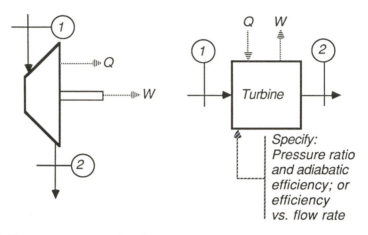

Figure 8.9 Flowsheet representation of a turbine.

8.5.3 Flow Splitter

Whenever a flow stream is to be split, intuition might indicate that this is really not a process. However, for the flowsheeting scheme described here, this has to be considered a process and given a separate block (or module). Figure 8.10 illustrates a splitter. A key parameter that must be specified is the split factor, $sf = m_2/m_1$. Since no work is associated with a simple splitting process, none is shown in the diagram. Also, since this is probably accomplished with a tee in the piping, the area for possible heat transfer is small, and any resulting heat transfer is taken as being negligible. Information on pressure drop should be included, however.

8.5.4 Heat Exchanger

When encountering heat exchangers, a very large set of possibilities exists for specifications. See Figure 8.11. Generally, however, two pieces of information are necessary regarding the heat transfer and the pressure drop. Usually included in the first are the overall heat transfer coefficient (set by the fluids used and their state) and any one of the following items: heat transfer area, minimum ΔT, or overall duty (heat transferred).

182

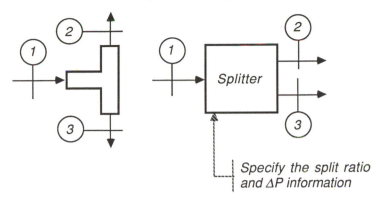

Figure 8.10 Block representation of a flow stream splitter.

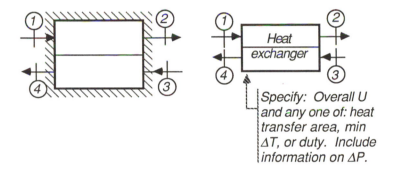

Figure 8.11 Schematic diagram of a heat exchanger, as well as the specification form for flowsheeting.

When the area and surface coefficient are set, this is the equivalent of fully defining the heat exchanger so that inputs of the temperatures and mass flow rates of the two incoming streams will allow the definition of all of the outputs from the block. Design specifications (as discussed earlier in this section) involve specification of the surface coefficient and either the minimum temperature difference or total heat transfer (duty). One possible way of solving for the outputs from a set of specifications of this sort is to estimate an area, find the resulting output, and compare this to the specification. A recycling through the calculations is needed until the specification and estimated values agree.

Do not forget to specify the fluid flow information that is pertinent to calculate the pressure drop through the device. This could involve any one of a number of specifications. Perhaps a vendor will furnish information on the pressure drop at a rated fluid flow rate for each of the streams. Then a ratio can be set up between the pressure drop and the square of the flow rate for a large range of flows.

REFERENCES

ASPEN, 1982, "Computer-Aided Industrial Process Design, the ASPEN Project, Final Report for the Period June 1, 1976 to November 30, 1981," U. S. Department of Energy Report MIT-2295T9-18, February 1.

Chinneck, J., and M. Chandrashekar, 1984, "Models of Large-Scale Industrial Energy Systems--I. Simulation," Energy, 9, pp. 21-34.

Curtis, B., M. Sood, and G. Reklaitis, 1981a, "Computer Aided Flowsheet Drawing--I. Equipment Layout," Computers and Chemical Engineering, 5, pp. 277-287.

Curtis, B., M. Sood, and G. Reklaitis, 1981b, "Computer Aided Flowsheet Drawing--II. Stream Layout and Drawing," Computers and Chemical Engineering, 5, pp. 289-298.

Evans, L., 1980, "Advances in Process Flowsheeting Systems," Chapter in FOUNDATION OF COMPUTER-AIDED CHEMICAL PROCESS DESIGN, VOLUME 1 (R. Mah and W. Seider, Eds.), Engineering Foundation, New York, pp. 425-469.

Hlavacek, V., 1977, "Analysis of a Complex Plant--Steady State and Transient Behavior," Computers and Chemical Engineering, 1, pp. 75-100.

Motard, R., M. Shacham, and E. Rosen, 1975, "Steady-State Chemical Process Simulation," AIChE Journal, 21, pp. 417-436.

O'Shima, E., 1980, "Static and Dynamic Simulation Programs of Japan," Chapter in FOUNDATION OF COMPUTER-AIDED CHEMICAL PROCESS DESIGN, VOLUME 1 (R. Mah and W. Seider, Eds.), Engineering Foundation, New York, pp. 511-527.

Roe, P., 1966, NETWORKS AND SYSTEMS, Addison-Wesley, Reading, Mass.

Rosen, E., 1962, "A Machine Computation Method for Performing Material Balances," Chemical Engineering Progress, 58, pp. 69-73.

Rosen, E., 1980, "Steady State Chemical Process Simulation: A State-of-the-Art Review," Chapter in COMPUTER APPLICATIONS IN CHEMICAL ENGINEERING, PROCESS DESIGN AND SIMULATION (R. Squires and G. Reklaitis, Eds.), ACS Symposium Series 124, American Chemical Society, Washington, DC.

Rosen, E., and A. Pauls, 1977, "Computer Aided Chemical Process Design: The FLOWTRAN System," Computers and Chemical Engineering, 1, pp. 11-21.

Timar, L., et al., 1984, "Useful Combination of the Sequential and Simultaneous Modular Strategy in a Flowsheeting Programme," Computer and Chemical Engineering, 8, pp. 185-194.

Westerberg, A., H. Hutchison, R. Motard, and P. Winter, 1979, PROCESS FLOWSHEETING, Cambridge University Press, Cambridge.

PROBLEMS

8.1 Consider the heat exchanger shown in Figure 8.12. Assume that this is a simple counterflow device. Also assume that the following information

184

is specified.

$$m_1, \ T_1, \ m_3, \ T_3, \ U \ \text{(Overall Heat Transfer Coefficient)}$$

Give expressions for the outgoing temperatures for each of the following cases.
(a) In addition to the previously noted information, the surface area is given.
(b) Instead of the surface area, assume that the minimum temperature difference, ΔT_{min}, is specified along with the other data noted above.

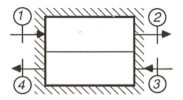

Figure 8.12 The heat exchanger system considered in Problem 8.1.

8.2 A turbine model is to be used in a simulation. Assume a system as shown in Figure 8.13. The working fluid is air at moderate pressures, and the incoming flow rate, temperature, and pressure, as well as the pressure ratio, are specified as m, T_1, P_1, and PR; but several different values of m are going to be used in the simulation. From a manufacturer it is known that the adiabatic efficiency of the device varies with incoming flow as follows for a range of flows $m_{min} < m < m_{max}$:

$$\eta = \beta_0 + \beta_1 \, m + \beta_2 \, m^2, \ \text{where the } \beta \text{ are constants}$$

Define completely the expressions that are used to find the outgoing values of pressure, temperature, and power.

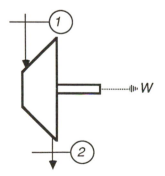

Figure 8.13 System discussed in Problem 8.2.

185

8.3 Set up the system flowsheet for the Brayton cycle shown in Figure 8.1. Assume that the compressor and the turbine each have a specified, constant efficiency of 0.82. Treat the combustor such that a maximum temperature can be achieved at the exit as a result of the heating process. Show the effect of maximum temperature on cycle efficiency. Solve the problem using the sequential-modular approach. Assume air is the working fluid and that it can be treated as an ideal gas with constant specific heats. You may use a different component/state-point notation than that given in Figure 8.1, but be sure to define any you use.

8.4 Repeat Problem 8.3 using a simultaneous-solution approach.

8.5 Using the system shown in Figure 8.14, set up the equations necessary to solve the problem simultaneously to find the unknown counterflow heat exchanger areas from the given information. Find one set of areas. Assume that the t_1 stream at 110°F has a mass-flow-specific-heat product of 2000 Btu/hr°F; the t_4 stream at 100°F has the value of this same product of 2500; the T_1 value is 1000°F; and the T_4 value is 1500°F. You may also assume that $T_3 = 150°F$, $T_6 = 240°F$, $t_3 = 410°F$, and $t_6 = 340°F$.

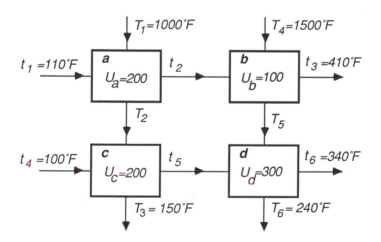

Figure 8.14 A system of interconnected heat exchangers to be used in solving Problem 8.5.

8.6 Define the system flowsheet for a Rankine cycle with one closed feedwater heater and no reheat. Assume that constant efficiencies are given for the pump and turbine. Indicate which terms must be defined by the user in solving for the various unknowns. Where there are choices between variables, so indicate.

8.7 Repeat Problem 8.6 for three feedwater heaters, the second being of the open type and the other two being closed.

186

CHAPTER 9

INTRODUCTION TO
OPTIMIZATION TECHNIQUES

9.1 INTRODUCTION

An *optimum* is defined by most dictionaries as being the best or most favorable of a set of choices. Human life is filled with a desire to optimize many things, including the use of time and money.

Two important steps are necessary in any optimization process. First, the process to be optimized has to be described; and in most human endeavors, the self-motivation for many things is not well understood by the individual, nor is the desired end result enunciated very well. For example, this involves how we spend any "extra" money we may have after we pay the rent and buy the food. Some people may have a critical need to save, or to invest, or to buy, or to gift. This may be motivated to build a nest egg, to "keep up with Joneses," or a variety of other stimuli. In terms that will be used later in this chapter, this overall strategy is called an *objective function*. That is, what is desired to be optimized? A key companion to the definition of the objective function is an understanding of any *constraints*. For example, consider that someone may wish to buy a new sports car each year that has a net cost (the difference between the cost of the new and the trade-in of the old) that is twice the yearly salary of that person. With no other source of money, this could be a real constraint. Large amounts of savings, or a spouse's salary, could modify the constraint situation. Together, though, the objective function and the constraints define a desired optimization.

Second, after the optimization problem has been defined, the next step is the *solution*. For example, how are you going to save the money for that new sports car and still invest some money, pay for the kids' education, or whatever else may be desirable?

Problems in thermal system design also often have two steps. Up until now, concepts have been discussed that can be useful in defining an objective function and corresponding constraints. If, for example, a steam pipe is to be insulated, the motivation often is to minimize total costs. That is, the cost of the insulation and its installation should be less than the savings of the cost of the energy lost. Heat transfer principles and cost information can be used in defining the overall cost function that is to be minimized. Any constraints must also be considered in the first step. The insulation will probably be manufactured only in specific sizes, obviously including some maximum thickness. There could also be some installation limitations that might restrict insulation to sizes less than the maximum available. The physical situation could be that if the steam drops below some specific energy content, it is unusable. All of these restrictions (constraints) could be described mathematically. These, along with the relationship for the overall cost that is to be minimized, describe the optimization problem.

The question addressed in the sections that follow is how to solve the optimization problem once it is defined. Finding the solution(s) for one of these situations can be quite complex and involved. A variety of techniques can (and sometimes must) be used to seek optimal values of the objective function.

Techniques described here will be kept quite limited in number, simple to use, and general in applicability. The price that is paid for this approach is that the techniques may not be very efficient. In fact, there might be times when it may appear that no solution will be found. However, the solution types described here will be a good basis to build an understanding of more complex and appropriate techniques.

9.2 GENERAL PROBLEM FORMALISM AND BASIC IDEAS

The general optimization problem can be stated as follows:

Minimize $F(X)$

Subject to $g_i(X) \leq 0, \ i = 1, 2, ..., m$ (9.1)

$h_j(X) = 0, \ j = 1, 2, ..., n$

Note: although optimization can mean either *minimization* or *maximization*, this will always be taken in what follows in terms of a minimization. This is done without lack of generalization because any maximization problem can always be reformulated in terms of a minimization.

In the notation used here, X denotes the vectorial representation of all of the independent variables in the problem. If this happened to be a problem that involved three-dimensional Cartesian coordinate system directions as the independent variables, X would be taken to denote variation with x, y, and z. If this were a thermodynamic system problem, X could denote the pressure, temperature, enthalpy, entropy, and any other independent variables in the problem. The $g(X)$ and $h(X)$ functions are known generally as the *constraints* on the problem. Constraints are of two types: inequality (denoted here by g) and equality (denoted here by h). Some simple examples of unconstrained and constrained <u>optimization</u> problems are shown in Figure 9.1.

In each case given in Figure 9.1, it is assumed that the minimum of the function *F(x)* (note that *x* is used here rather than *X* because a function of a single variable is represented) is to be found in the region shown. The unconstrained problem is shown in the top portion of this figure. With no $g_i(x)$ or $h_j(x)$ specified, the minimum of the function within the region of interest is the optimal value of the objective function.

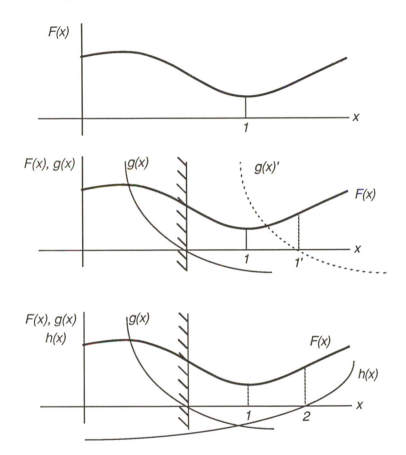

Figure 9.1 A sketch showing several simple situations for various optimization problems. For a description of the aspects shown, see the accompanying text.

The middle plot in Figure 9.1 indicates the effects of an inequality constraint on the solution. Here it is assumed that the system to be solved is simplified from the set of Equation 9.1 to the following: minimize *F(x)* subject to $g(x) \leq 0$. An inequality constraint such as *g(x)* here defines *feasible* and *infeasible* regions. In the second plot in Figure 9.1, the feasible region is the area to the right of the crosshatched vertical line. As shown there, the particular *g(x)* given does not change the solution of the problem compared to the unconstrained situation discussed earlier. If, however, the constraint function had crossed the axis more to the right, as shown by dashed lines, the solution would have been at the point denoted as *1'* rather than the original solution denoted as *1*.

Finally, consider the third portion of Figure 9.1. Here the same problem as was considered above is repeated with one additional constraint added. This is an equality constraint *h(x)*. Now the optimization problem is: minimize *F(x)* subject to *g(x)* ≤ 0 and *h(x)* = 0. As shown in the figure, *h(x)* = 0 occurs at only one value of *x*. This point is in a feasible region as defined by *g(x)*, so the solution is located by this point (denoted by *2*). It is frequently the case that all of the constraints cannot be satisfied at the same time. This is particularly true if the problem of interest has a great number of constraints. For example, *h(x)* may have crossed the *x* axis to the left of the crosshatched area. In this case, the problem would not have a completely satisfied solution. The designer then has the option of abandoning the problem, reformulating the problem, or finding the least unsatisfactory solution.

There will be situations when a minimization of a function will be found analytically (although most times this will be done with at least some element of numerical solution). As is well known to any student of calculus, the necessary condition for a minimum of a function to occur is to have the first derivative of the function vanish. That is,

$$d\,F(x)\,/\,dx = 0 \qquad\qquad (9.2)$$

or in general *n* space:

$$\nabla \cdot F(\mathbf{X}) = 0 \qquad\qquad (9.2a)$$

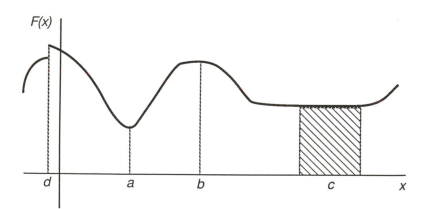

Figure 9.2 Plot of a function that shows several stationary points and one discontinuity.

While the requirement on the derivative vanishing is necessary, it is not sufficient, as can be seen in Figure 9.2. Of course, the value of the second derivative can help greatly in determining which of these situations is the desired minimum, but often that information may not be available.

Points *a*, *b*, and *c* all satisfy Equation 9.2. However, there is only one absolute minimum for the region shown, and that is point *a*. As will be shown later, methods that find the minimum by comparing values of the function at a variety of locations in the region of interest will not have a problem with the behavior shown. However, as will also be shown, it is desirable to make optimization techniques more efficient by eliminating usually unnecessary calculations. In this case, the technique might "home in" on local minimum without checking if it is really the absolute minimum.

Another problem arises with some techniques when the function of interest is discontinuous. An example of this occurs at point *d* in Figure 9.2. Some techniques also have problems with situations when the first derivative is discontinuous. This will appear on a plot as a sharp change in direction, but no gap, in the function.

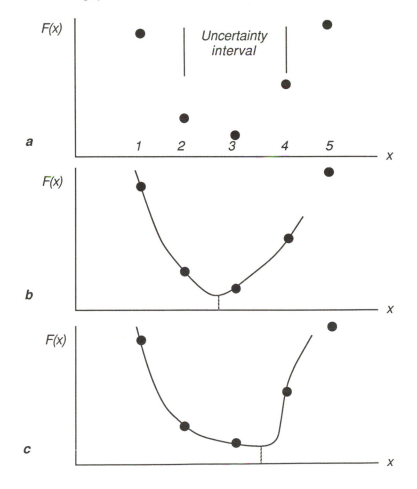

Figure 9.3 Demonstration of what is meant by the interval of uncertainty. For the function shown, this interval lies between points *2* and *4*.

The concept of *uncertainty interval* should be noted. Initially, it is the total region of interest. At the end of the search, it encompasses three trial values of the independent parameter. To visualize this latter statement, consider the function shown in Figure 9.3*a*, which is being plotted from discreet values of the independent variable. There is an uncertainty where the minimum of the function lies based upon the five values shown. As shown in Figure 9.3*b*, the minimum could be between points *2* and *3*; but as shown in Figure 9.3*c*, it could be between points *3* and *4*. From the evaluations shown, the minimum is between points *2* and *4*; and this is the interval of uncertainty.

A necessary functional trait in many solution approaches is that of *unimodality.* This term means that there is only one minimum in the functional region of interest. Values of *x* closer to the minimum of the function will yield smaller values of the function than values of *x* farther away from the minimum. This statement holds for a comparison of functional values on the same side of the minimum. In Figure 9.3, $F(x_1) >$ $F(x_2)$ and $F(x_5) > F(x_4)$. If this type of relationship holds for each pair of points that is on one side of the minimum, the function is considered to be unimodal.

9.3 UNCONSTRAINED FUNCTIONS OF ONE INDEPENDENT VARIABLE

In the consideration of the design analysis of thermal systems, functions of one variable are simultaneously not very important and extremely important. The reasons for this seemingly conflicting statement are twofold. First, it is very seldom that the design involves a problem of only one independent variable. On the other hand, problems with a single variable arise many times at intermediate points in the solution of more involved formulations. Some multivariable problems use a single-variable technique as a portion of the total solution.

For the following discussion, assume that the functions of interest are unimodal. While most are in fact multimodal, often it is not too difficult to reduce the problem to a single unimodal region of interest (Sargent, 1973; Vanderplaats, 1984). Typically, this might involve a coarse grid evaluation of the function, reducing the overall region to the corresponding unimodal uncertainty interval.

Consider a qualitative discussion of techniques for finding the minimum of a unimodal function. The set of approaches noted here is necessarily brief and incomplete, but these are often valuable techniques to use.

9.3.1 Sequential or Exhaustive Search

In the exhaustive search technique, evenly spaced, discreet evaluations are made for the function, starting at one end of the region of interest and marching to the other. A comparison of all values will show where the uncertainty interval lies for the function's minimum. It would usually be the case that the uncertainty interval from the first-pass evaluation will become the region of interest for a second pass. An increasing number of consecutively more focused passes can thus be configured, achieving any small final uncertainty interval desired.

Normally, the assumed unimodal feature of the function is exploited to decrease the total number of function evaluations. In this approach, a sequential set of evaluations is made from one end of the region of interest. Because of unimodality, the functional values will decrease continuously until one value will be larger than the previously evaluated one. This value of x and the second previous value of x define the uncertainty interval. Further definition of the minimum can follow in a similar vein.

9.3.2 Golden Section Search

The Golden Section Search takes advantage of information developed in previous steps of a minimization while continually eliminating a large portion of the search region in each further evaluation step. As before, the unimodal characteristic assumed for the region is used to benefit in the analysis.

A prime concern is that symmetry be exhibited in the search pattern. For a normalized overall interval distance of unity, take each trial to be located a fraction v from each end. On each elimination, remove the amount $1-v$ (i.e., leave the amount v). The next trial should be located at a fraction v, as before. Thus, it is necessary to solve for v from the relationship $1-v = v^2$. Find $v = 0.61803. \ldots$ It should be noted in passing that this same result can be shown to be related to the limit of two consecutive, large *Fibonacci numbers*. Several authors give good theoretical developments of the Golden Section Search and the more fundamental Fibonacci Search. See, for example, Gill et al. (1981).

Now apply this idea. Locate one point $38.197. \ldots\%$ across the region. Call that location x_1. Evaluate the function again at a location $61.803. \ldots\%$ of the distance across the region. Note that the two points are located the same distance from the centerline of the region of uncertainty. Denote that location as x_2. Now compare $F_1 = F(x_1)$ with $F_2 = F(x_2)$. If $F_2 > F_1$, then it is clear that the minimum of the function does not lie in the region between x_2 and x_{max}. (Keep in mind that this conclusion can be drawn because the function is unimodal.) If $F_2 < F_1$, then the region below x_1 can be eliminated.

Now a third evaluation of the function is located symmetrically with the remaining point inside the uncertainty interval. Elimination of one of the regions proceeds as before. The general approach is shown schematically in Figure 9.4, and the approach is described in the discussion that follows.

Consider the calculation of a minimum of a function, as shown in Figure 9.4. The original uncertainty interval is assumed to be between the two sets of crosshatched marks. Two calculations of the function are made: one (numbered *1* in the figure) at a location 38.197. . .% across the interval, and another (numbered *2* in the figure) at a location 61.803. . .% of the total interval. Since the functional value at *1* is greater than the corresponding value at *2*, and since the function is assumed unimodal, then the minimum must lie between *1* and the upper limit of the complete interval. Hence, the region to the left of *1* can be eliminated from further consideration. This is shown in the second part of the figure, where a new left limit of the region is shown.

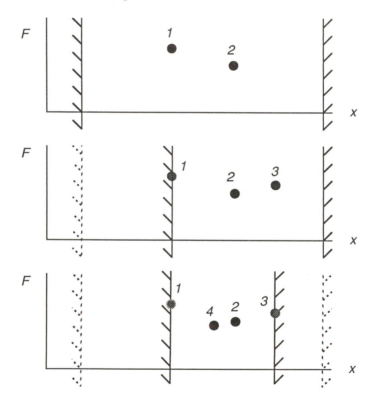

Figure 9.4 The basic idea of a Golden Section Minimization Search, shown for four evaluations. After the fourth calculation of the function, the uncertainty interval is found to lie between points *1* and *2*. Over 3/4 of the original interval has been eliminated at this point.

A third calculation of the function is made such that the x location is symmetrical with x_2 in the remaining uncertainty interval. x_3 is located at a location 77.394% across the original interval. If F_3 is larger than F_2, as shown, then the region to the right of x_3 can be eliminated. This is shown in the third portion of the figure. A fourth trial is then calculated at

x_4 (symmetrical with x_2 in the remaining interval) and the function yielded a value lower than any others. If the calculation is halted at this point, the uncertainty interval is between x_1 and x_2, a decrease to approximately 1/4 of the original region of interest.

9.3.3 Polynomial Approximations

The basis of this approach is to evaluate functional values, and, possibly, functional-slope values at a number of points, and fit a polynomial to the data. (See Chapter 4 on curve fitting.) A minimum can then be found for the polynomial via analytical techniques. The location of the estimated minimum is then used as a starting point for a second set of curve fitting and minimum point estimation. Generally, a quadratic equation is used for the curve fit, although cubic equations sometimes yield desirable results. Higher order functions are more complicated to program and typically do not represent well the simple curve forms needed for this approach. Further information on the polynomial approximation method is given in a number of sources. See, for example, Sargent (1973), Gill et al. (1981), Reklaitis et al. (1983), and Vanderplaats (1984).

9.4 CONSTRAINED MINIMIZATION OF FUNCTIONS OF A SINGLE VARIABLE

Constraints are present in even the most elementary problems in design. Often the constraints are not considered in any special formal way, but, instead, are handled informally. Such needs as requiring flow to be in one direction only or a difference in temperature to be always positive are examples of inequality constraints. Equality constraints can arise from property and heat transfer correlations, to name only two classes of a very large set of examples.

Consider an optimization of the following form.

Objective function:

$$F(x) \quad = \quad \frac{1}{|x - \Omega|} \quad + \quad \frac{1}{|x - \Delta|} \tag{9.3}$$

Equality constraint:

$$h(x) = \mu + \phi x = 0$$

where Ω, Δ, μ, and ϕ are constants. A possible form of these equations might appear as shown in Figure 9.5. The unconstrained minimum of the $F(x)$ function, only, would be at the location denoted as a in that figure. Ignoring the fact that this particular problem can be handled quite easily in an analytical form (i.e., solve the constraint for the pertinent value of x and substitute this value into the objective function), how can this minimization be performed in a general way?

A way that constraints are often handled in optimization problems is to construct another function that combines the objective function with a pertinent representation of the constraints. The result is a single equation that minimizes in a fashion similar to nonconstrained problems but reflects the effect(s) of the constraint(s). This approach is known by a variety of names, including *compound function* (Lootsma, 1973), *psuedo-objective function* (Vanderplaats, 1984), *penalty* or *barrier function* (two slightly different concepts) (Gill et al., 1981; Reklaitis et al., 1983), and *generalized penalty function* (Westerberg, 1981). While some of the critical details vary between these concepts, consider an illustrative example of the basic idea with regard to the constrained optimization defined above.

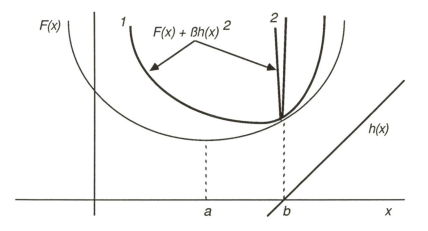

Figure 9.5 An example of the use of a penalty function in an equality-constrained optimization. The darker curves represent $0 < \beta_1 < \beta_2$.

From the system shown in Equations 9.3, combine the original function and the constraint into a single function $F(x)$ as follows:

$$F(x) = F(x) + \beta \, h(x)^2 \qquad (9.3a)$$

Depending upon the value of the *penalty parameter*, β, Equation 9.3a can represent a wide range of characteristics of Equations 9.3. Obviously, for $\beta=0$, $F(x) = F(x)$. Moderate values of β yield a slightly modified curve from the original (see curve *1* in Figure 9.5). Very large values of β will increasingly concentrate the curve about the constrained minimum. This is shown as curve *2* in Figure 9.5. Hence, the minimum of Equation 9.3a for large values of β will be the minimum of the original constrained system.

Inequality constraints can be handled in a similar manner. For example, had the system been as shown in Equations 9.3 except for the constraint statement given as $g(x) \leq 0$, a penalty function could still be

196

defined in the same manner as was done in Equation 9.3a. Or:

$$F(x) = F(x) + \beta\, g(x)^2 \qquad (9.4)$$

Now, however, β would be given some value (as before) only when $g(x) > 0$. When the inequality constraint is met, β would be given the value of zero. In the particular case of the functions shown in Figure 9.5 (assuming the form of the inequality constraint is the same as that of the equality contraint), the constrained optimum with an inequality constraint would be the same as the initial unconstrained problem solution.

This all seems quite straightforward, and it is; but the approach is not without its problems. If, for example, the function $F(x)$ is evaluated at some distance from the optimum value with a large value of β, extremely large numbers (perhaps computer overflow conditions) will result. Another problem is that sharp discontinuities in $F(x)$ develop near the optimum point. Some minimization search techniques that require the function be smooth, like the polynomial approximation, can be adversely affected by this characteristic. Techniques that merge ideas from Lagrange multipliers (not covered here) and other concepts with penalty functions circumvent these types of difficulties (Lootsma, 1973; Gill et al., 1981; Westerberg, 1981).

9.5 UNCONSTRAINED OPTIMIZATION OF MULTIVARIABLE FUNCTIONS

In moving into a discussion of problems that have more than one independent variable, the situation becomes simultaneously more realistic and much more complex. Here, the problem is truly as stated in Equation 9.1. X represents n variables, such as $x_1, x_2, \ldots, x_n$. Functions of two variables can be shown quite easily in a two-dimensional plot via the use of contour lines, and use of this ability should be made to assist in the visualization of physical situations whenever needed. Functions of more than two variables become much more difficult to depict in a two-dimensional plot. If desired in these cases, either special visualization methods have to be used (e.g., taking the independent variables two at a time), or the results of the evaluation must be inspected in strictly numerical form.

Consider the minimization of the function $F(x_1, x_2)$ with no constraints. Suppose this function plots as is shown in Figure 9.6, where contours are shown with larger values of the curve index in the figure. That is, $F_1 > F_2 > \ldots > F_5$. The minimum value of the function in the region shown is assumed to be located inside the F_5 region. For this bowl-like shape, the challenge of a minimization technique is, starting at any location, to find the coordinates of the minimum value of the function. By analogy to skiing, the quickest way to the bottom of the slope is straight down, perpendicular locally to the contours. A technique that uses this idea is discussed later.

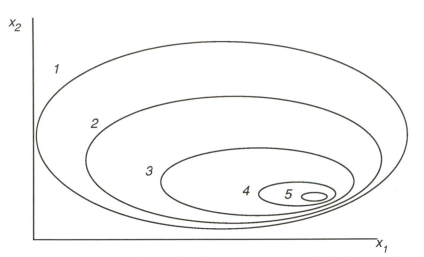

Figure 9.6 A representation of a function of two independent variables by use of contour lines.

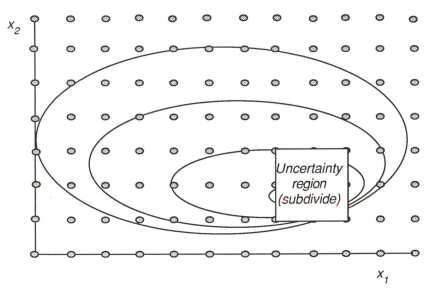
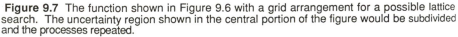

Figure 9.7 The function shown in Figure 9.6 with a grid arrangement for a possible lattice search. The uncertainty region shown in the central portion of the figure would be subdivided and the processes repeated.

Numerous methods are available for finding a minimum value of a function of several variables without constraints. Consider some simple, but possibly very labor-intensive and far-from-foolproof, techniques for doing this.

One of the easier techniques to visualize is the "brute force" method of a lattice search. This approach involves the evaluation of the function at regular spacings of the independent variables, taken in combination. In Figure 9.6, this could be represented as functional computations at values of the following: Δx_1, $2\,\Delta x_1$, $3\,\Delta x_1$, $4\,\Delta x_1$, . . ., each paired with each of the following: Δx_2, $2\,\Delta x_2$, $3\,\Delta x_2$, $4\,\Delta x_2$, . . . Thus, there would be regularly spaced evaluations throughout the initial region of uncertainty. The lattice that might be used is shown as a series of gray dots in Figure 9.7. A resulting region of uncertainty could then be defined to within 3^n (where n is the number of independent variables) grid locations. Remember that the uncertainty interval for a single-variable evaluation shown in Figure 9.3 involved three grid points and a $2\,\Delta x$ spacing. For the two-variable situation shown in Figure 9.6, the uncertainty region would involve nine grid points and a $2\,\Delta x_1$ by $2\,\Delta x_2$ spacing. Usually, the first step of the search is performed on a coarse grid arrangement. The resulting region of uncertainty is then subdivided, and the whole process is repeated.

Clearly there is a trade-off between taking only a few, widely spaced evaluations on the first step or taking a large number of more closely spaced evaluations. The computation effort clearly can be very large in the latter situation. However, the former situation may not discover some localized minima. This technique tends to eliminate complicated programming problems, can often be used to handle multimodal functions, but involves a very large number of calculations. Many authors have reported the use of this technique. See, for example, the use of this general approach for cooling systems (Crozier, 1980), heat exchangers (Mott et al., 1972; Andeen and Glicksman, 1972; Kroger, 1983), solar collector parameters (Garg et al., 1981), and power plants (McConnell and Elmenius, 1976; Rikhter et al., 1976).

There is a more efficient variation on the brute force approach just described. Instead of evaluating the whole field in evenly spaced increments, simply evaluate an array of points in some local region. (One example is to use a three-point, or equilateral triangular, set in a two-variable function. This is usually termed the *Simplex Search* (Reklaitis et al., 1983).) By evaluating the values of the function at all of the points, a direction for movement can then be defined. Usually, the "moving" solution takes place so that a fair number of previously found values of the function can be used at each step.

A second approach to multivariable minimization is to fix $n - 1$ of the variables and minimize the remaining variable. Fix this variable with all but one of the originally fixed variables, and perform a minimization on the one remaining variable. Cycle through the complete set in this manner, and then start over. This can be called a *Series-Univariant* optimization. There are convergence problems with this method for a variety of functions. For the form shown in Figure 9.6, however, the procedure would perform quite satisfactorily.

An example of a *Series-Univariant Search* is shown in Figure 9.8, where each of the steps *1*, *2*, and *3* are shown both on the main graph and in detail below. It is not unreasonable to assume that the one-dimensional grid spacing similar to that shown in Figure 9.7 could be used. Hence, it seems obvious that the minimum could be found in many fewer calculations than a lattice search. However, keep in mind that there is a one-dimensional minimization for each step shown.

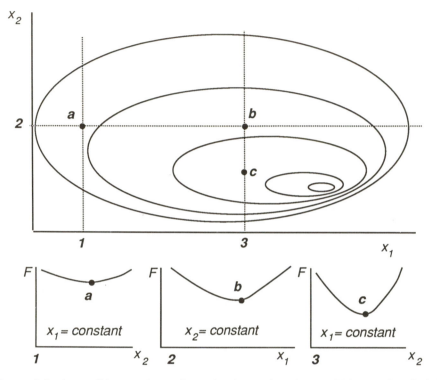

Figure 9.8 A possible set of one-dimensional searches that could be used to find the minimum of the function of two variables shown in Figure 9.6. Do not be misled by the seemingly efficient operation of this technique. Most functions will not yield this easily!

More efficient approaches have been used that are patterned after the concept shown in Figure 9.8. The idea is to start out in the same manner as just described; but after the third step, modify the way directions are chosen. Define the fourth search direction from the second and third. One way of doing this is shown in Figure 9.9. Discussion of many of these techniques is given by Sargent (1973).

The concepts that have been discussed up to this point simply use values of the function to attempt to define the minimum of the function. While these tend to be straightforward to illustrate and program, in fact, the techniques are generally computationally intensive and often slow to converge.

Another category of solution techniques involves the use of the gradient of the function. Analogously, skiing or climbing down a mountain, the path that has the greatest slope will generally lead to the bottom of the mountain the fastest. These approaches are often termed the methods of "steepest descent." In concept, the idea behind the steepest descent approaches is quite straightforward. At the current location, it is simply necessary to evaluate the gradient of the function. That is, find $\nabla F(X)$, and move in the direction opposite to this. Depending upon the form of the function, convergence can be quite slow, and this has been discussed by a number of authors. (See, e.g., Gill, et al., 1981; Vanderplaats, 1984.) An example of a very efficient use of the technique is shown in Figure 9.10.

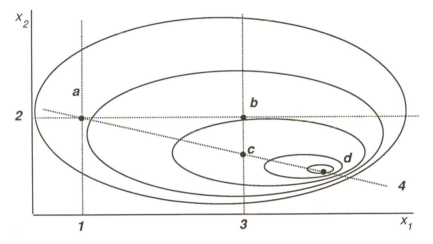

Figure 9.9 Use of the first three univariant optimizations to define the direction of the fourth. A variety of approaches use some aspect like this to yield a more efficient computation sequence.

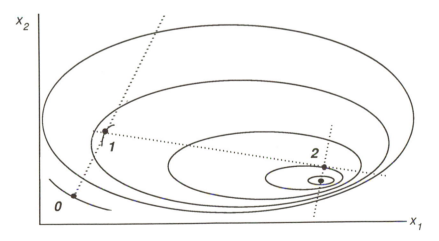

Figure 9.10 An example of use of the method of steepest descent. At each location, a new direction is set by moving toward the negative of the gradient of the function.

In many cases, simple modifications of method of steepest descent vastly improve the convergence and/or the use of computer storage. One such technique is discussed by Fletcher and Reeves (1964). This approach will not be illustrated here, but it is related to a combination of the techniques shown in Figures 9.9 and 9.10. In all cases using these techniques, though, the gradient must be calculated at each evaluation point. It is not hard to imagine that roundoff errors could generate computational difficulties in some complicated functions.

One final category of unconstrained optimization techniques will be noted. This category invokes the use of second derivatives. Since functions of several variables are being dealt with here, this gives rise to a matrix called the *Hessian*. This can be written as in Equation 9.5.

$$[H(X)] = \begin{bmatrix} \dfrac{\partial^2 F(X)}{\partial X_1{}^2} & \dfrac{\partial^2 F(X)}{\partial X_1 \partial X_2} & \cdots & \dfrac{\partial^2 F(X)}{\partial X_1 \partial X_n} \\[2ex] \dfrac{\partial^2 F(X)}{\partial X_2 \partial X_1} & \dfrac{\partial^2 F(X)}{\partial X_2{}^2} & \cdots & \dfrac{\partial^2 F(X)}{\partial X_2 \partial X_n} \\[2ex] \cdots & \cdots & \cdots & \cdots \\[2ex] \dfrac{\partial^2 F(X)}{\partial X_n \partial X_1} & \dfrac{\partial^2 F(X)}{\partial X_n \partial X_2} & \cdots & \dfrac{\partial^2 F(X)}{\partial X_n{}^2} \end{bmatrix} \qquad (9.5)$$

To see what role this plays in optimization solvers, consider a simple function of one variable. While this is obviously a trivial case, it will yield a more transparent explanation than the more complete formulation.

Consider a Taylor series expansion of the function $F(X)$ about the point X_o.

$$F(X) = F(X_o) + F'(X_o)\, \delta X + (1/2) F''(X_o)\, \delta X^2 + \dots \qquad (9.6)$$

In the notation used here $\delta X \equiv X - X_o$. Now differentiate the Taylor's expansion expression with respect to X. The following results:

$$F'(X) = 0 + F'(X_o) + F''(X_o)\, \delta X + \dots \qquad (9.7)$$

Assume that X is the location of the minimum; so it is a necessary condition that $F'(X) = 0$. Solving this equation for the location X, the following is the result:

$$X = X_o - F''(X_o)^{-1}\, F'(X_o) \qquad (9.8)$$

Generalizing this to n variables by using the Hessian, the comparable result to the single-variable equation is as follows:

$$X = X_o - H(X_o)^{-1}\, \nabla F(X) \qquad (9.9)$$

The value of the Hessian of a function at a given point in n space has implications about the form of the function. This will become more apparent in the following discussion.

Calculation of the Hessian is the basis for a class of search methods. In a manner similar to the approach used in the gradient method, the second derivative can be used in a step-by-step search for the minimum of the objective function. Repeating how a function is expanded in a Taylor series (above) but for the general n-space situation, the function can be written as shown in Equation 9.10. Here it is assumed that the series calculation is performed in a stepwise manner, and the superscript i denotes the ith iteration.

$$F(X) \approx F(X^i) + \nabla F(X^i)^T \cdot \delta X + (1/2)\,\delta X^T \cdot H(X^i)\,\delta X \qquad (9.10)$$

Here the additional notation of $\delta X = X^{i+1} - X^i$ is used. This can be used to estimate the value of the independent parameters that will result in the stationary point. Solving Equation 9.10 for the stationary conditions (i.e., the gradient of the function is zero),

$$X^{i+1} = X^i - H(X^i)^{-1}\,\nabla F(X^i) \qquad (9.11)$$

which is, of course, the same result as shown in Equation 9.9. This approach is most often known as *Newton's method*.

While several benefits are associated with speed of convergence of a Newton's method approach, the technique is not without its practical problems. Not the least of these is the potential of having a singular Hessian. It is obvious from Equation 9.11 that even a nearly singular Hessian can result in large changes in the predicted values of the stationary point. Hence, essentially unstable solutions can often result from the use of these techniques unless special safeguards are built into the approach. A number of variations can simplify the general evaluation process if certain characteristics are present in the function being evaluated. Many of these topics have been examined extensively in the literature. (See, e.g., Sargent, 1973; Gill et al., 1981; Reklaitis et al., 1983; Vanderplaats, 1984.)

9.6 MULTIVARIABLE OPTIMIZATION WITH CONSTRAINTS

9.6.1 Linear Programming

Linear programming (LP) is important for several reasons. First, LP is by far the most used approach to optimization at the present time. Second, some problems that are nonlinear can be cast into a linear form and attacked by LP. Third, a comprehensive understanding of LP can aid in having a better grasp of nonlinear problem solution techniques and the computational problems of nonlinear solution techniques. Having said all of that, however, the overriding fact remains that few problems in the design of thermal systems are easily amenable to LP approaches. The discussion of this technique is included here only because of the tremendous importance of LP in the optimization field generally. Not to mention it in an introduction to optimization would be a serious omission.

Usually, an LP problem is expressed in much the same way as the general optimization statement given in Equation 9.1. Here, however, there is always a linear relationship between the variables. Thus, the objective function and constraints are usually stated as:

Minimize
$$F(\textbf{X}) = \sum_{i=1}^{n} c_i x_i$$

Subject to
$$\sum_{i=1}^{n} a_{ij} x_i = b_j \qquad j= 1, 2, ..., m \qquad (9.12)$$

with $x_i \geq 0$ and $b_j \geq 0$

Any inequality constraints are cast into equality constraints by use of an ad hoc variable that either adds (in this case it is termed a *slack variable*) or subtracts (termed a *surplus variable*) to remove the inequality. (Similar techniques can be used in *nonlinear programming* methods.)

Consider a simplistic example of linear programming as might be applied to a cogeneration situation. Assume that the output of a steam plant can be either electrical power (E) or thermal power (T). Further assume that the value of electrical power is three times that of thermal power. The total value of "product" from the plant is to be maximized. Or:

Maximize
$$F = 3E + T \qquad (9.13)$$

If this were all that governed the problem, the solution would be to produce electricity only and, in addition, to produce a great deal of it! Consider additional limitations to the problem. Clearly, a given plant has some overall electrical generation capacity, say E_{max}, that is set by the size of the turbine/generators. For the sake of being specific, take the result shown in Equation 9.14.

$$E_{max} = 65 \qquad (9.14)$$

Another obvious limitation is that not all of the energy could ever be transformed completely to electricity. Even when there is no electricity produced, some of the boiler input does not show up as output because of losses. Relating the efficiencies of both the boiler and the remainder of the system, the total boiler capacity will limit the output in the manner shown in the expression in Equation 9.15.

$$E / \eta_{total} + T / \eta_{boiler} \leq Total\ boiler\ capacity \qquad (9.15)$$

or, using some arbitrary numbers to demonstrate the effects,

$$2.5E + 1.1T \leq 200 \qquad\qquad (9.15a)$$

The results of Equations 9.13-9.15a are shown in Figure 9.11. Note that the bold lines represent the equality cases. Hence, permissible solutions fall below the bold lines.

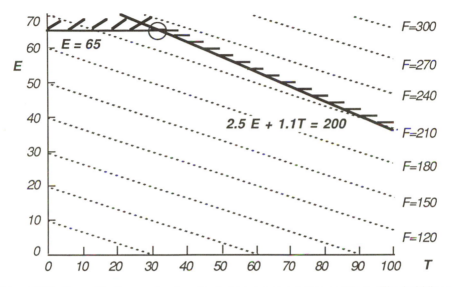

Figure 9.11 A graphical example of a simple linear programming problem. The bold lines represent the equality state of the constraints, while the dashed lines indicate constant values of the objective function.

For the problem shown in Figure 9.11, the solution is easily found by simply eliminating one of the two variables between the two constraints. When this is done, the location of the optimal value (the circled point in Figure 9.11) is found to be at $E = 65$ and $T = 34$. Substituting these values into Equation 9.13 yields $F = 229$.

It is not hard to imagine that if the number of variables and constraints grow, the direct-solution method used above would not be viable. For this reason, several formal techniques have been developed for the solution of LP problems. Perhaps the most used approach is the *simplex method*, which should not be mistaken with a similarly named technique used for the solution of nonlinear problems noted earlier in this discussion. The possible confusion that can result from these names is unfortunate.

An observation that can be made for the solution shown in Figure 9.11 is generally valid: the solution always lies on the *boundary* of the region defined by the constraints at a *vertex* formed by the intersection of two constraints. The simplex method for solving LP problems normally involves two steps. First, a vertex is found and then adjacent vertices are compared until either the optimal value of the objective function is found

205

or it is determined that the latter is not finite.

The existing theory related to LP is quite extensive. For further information see the numerous works on this subject (e.g., Dantzig, 1963; Dantzig et al., 1981; Ravindran, 1982). As noted earlier, linear programming has been applied to nonlinear problems. An example of this where a nonlinear heat exchanger system is linearized for analysis is given by Henley and Williams (1973). Several recent applications to linear systems have appeared (e.g., Nishio et al. 1985). A graphical solution for a cogeneration problem has been reported (Karamchetty, 1980).

9.6.2 Nonlinear Systems with Constraints

9.6.2.a Introduction

Techniques are available for solution to nonlinear optimization problems with nonlinear constraints. While some of the possibilities mentioned below are built from concepts discussed earlier in this chapter, some use ideas not developed here at all. One such group of techniques uses *Lagrange multipliers.* The latter and their basis in optimization have been discussed by a number of authors. (See, e.g., Peterson, 1973; Gill et al., 1981; Reklaitis et al., 1983.) Also of value in specific situations is *geometric programming.* (See, e.g., Avriel and Wilde, 1967; Beightler and Phillips, 1976; Paul, 1982.) Neither the Lagrange multipliers nor geometric programming will be discussed here.

Most of the nonlinear techniques, no matter their theoretical basis, have one of two general ways of handling constraints.

(i) Some approaches are based upon configuring the objective function so that the latter reflects the effects of any constraints present. The augmentation of the objective function with a penalty function is an example of this. The single-variable application of this approach has been discussed earlier in this chapter.

(ii) Other approaches use special tactics in setting the search direction whenever a constraint is encountered.

9.6.2.b Use of Barrier or Penalty Functions

By far the most important of the various techniques for solving the systems of interest here are based upon solution methods for unconstrained problems. One of the two most important classes of these problems is termed an *interior-point method* and involves the minimization of penalty or barrier functions (see previous discussion) of the form

$$\text{Minimize} \qquad F(\boldsymbol{X}) - r \sum_{i=1}^{m} \phi[\, g_i(\boldsymbol{X}) \,] \qquad\qquad (9.16)$$

Here r is a positive variable and $g_i(X)$ denotes the constraints. As noted by Lootsma (1973), there are three particular subcategories of interest here. Denoting ø as a function of only one variable as shown, say $ø[\Omega]$, the three forms most often used are $ø = ln\ \Omega$, which is called the *logarithmic programming method;* $ø = -1/\Omega$, which is the single most applied technique and is known as *sequential unconstrained minimization technique,* more simply known by its initials *SUMT;* and the inverse square function $ø = -1/\Omega^2$. All of these specific functional forms are such that $ø[\ 0+\ ] = -\infty$.

The second major set of techniques is called *exterior-point methods.* In a very similar form to Equation 9.16, the general problem is given by Equation 9.17.

Minimize
$$F(X) - s^{-1} \sum_{i=1}^{m} f[\ g_i(X)\ ] \qquad (9.17)$$

Here s is also a positive variable. The form of f is such that $f[\Omega] = 0$ for $\Omega \geq 0$ and $f[\Omega] < 0$ for $\Omega < 0$, and these are often called *loss functions.* Clearly, the loss functions contribute to the overall composite function if the X values are infeasible.

In all situations, the augmented objective function reflects inequality constraints as high ridges in the functional hyperspace, while equality constraints become valleys. Typically, the solution will be on one of the equality constraints. In the solution technique the steepness of any ridges is sequentially increased. At the same time, the valley(s) for the equality constraint(s) (if present) can be made steeper sequentially. See Figure 9.12. Often, the combination of equality and inequality constraints is expressed as shown in Equation 9.18.

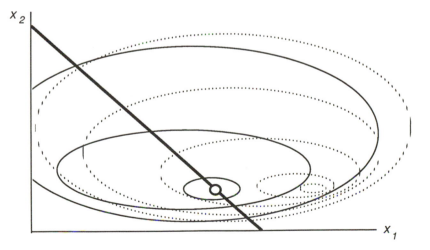

Figure 9.12 A mapping occurs when penalty functions are used in converting constrained problems (here a linear equality constraint is shown as the heavy line) into equivalent unconstrained problems. The dashed lines represent the original function as shown in Figure 9.6 and the solid curves indicate a possible transformed set of contours.

Minimize $\qquad$ $F(X) - \sum_i \alpha_i / g_i(X) + \sum_j \beta_j \, [\, h_j(X)\,]^2$ $\qquad$ (9.18)

In this objective function, both α and β are taken to be held constant during the various steps of a given search but varied after a minimum is found. Their values are increased sequentially. That is, values for the two constants are set; the minimization is carried out; the values of the constants are set to higher values; the minimization is carried out again; and the process continues until the location of the minimum changes sufficiently little that convergence can be assumed. The technique has several potential problem areas, including the fact that the rates at which the constants are increased need to be set with some insight and judgment. Also, when the valleys become sufficiently steep and the ridges sufficiently high, small movement steps may overshoot a valley completely without any outward indication, and/or computer overflows can occur.

A good survey of the area of constrained optimization has been given by Sargent (1980). Ray et al. (1971) have applied the SUMT technique to the optimization of a typical chemical process made up of several conventional components. This same process has been analyzed recently by Vinante and Valladares (1985) using a method based upon Lagrangian multipliers.

9.6.2.c Techniques That Control Search Direction and Step Size

There are a number of approaches that do not modify the objective function but, instead, affect the direction and step size in which each step in a search proceeds. Briefly these aspects are summarized below.

(i) Equality constraints are monitored; and when they are about to be crossed, the step size is decreased to allow the next step to fall on the constraint.

(ii) Inequality constraints affect the direction of the search in the proximity of the constraint. As an inequality constraint is approached, a new direction is set that is usually taken as some vector combination of the negative of the gradient vector of the objective function and the gradient vector of the constraint taken in the feasible direction. An example of this is shown in Figure 9.13.

9.6.2.d Other Techniques

A number of techniques use some combination of the methods described above or some fundamentally different approach. (See, e.g., Fletcher, 1981; Reklaitis et al., 1983.) Papers have appeared that have used other techniques (e.g., Hrymak et al., 1985).

9.7 PRACTICAL CONSIDERATIONS IN OPTIMIZATION SOLUTIONS

Many techniques require that the functions minimized have a unimodal form and continuous first few derivatives. If the first of these conditions is not true for an objective function of interest, then a solution found may not be the "best." From a practical standpoint, this may not be as bad as it might first appear. The solution found will undoubtedly be better than one that may be chosen without any analysis. On the second point, many problems arise with discontinuous first derivatives present in the simulation. Techniques are available for the treatment of these kinds of problems (e.g., Umeda and Ichikawa, 1971); but, as might be expected, the calculations increase greatly over similarly sized systems. Smooth first and second derivatives allow the use of very efficient techniques. Saddlepoints offer special problems. They may appear to yield a minimum in one direction while actually having a maximum in an orthogonal direction (Westerberg, 1981).

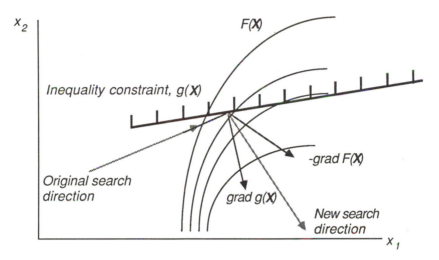

Figure 9.13 Pictorial representation of the determination of a possible change in direction that can take place when a search direction encounters an inequality constraint. The gradient of the objective function and the gradient of the constraint are added vectorially to set the new direction. Actually, the normalized gradients are usually used.

Constraints are found in virtually every problem. These are imposed not to allow a negative temperature difference in a heat exchanger or an unreal flow reversal. Often, the equality constraints are used to find new locations in the search. If a problem has inequality constraints, one of these will very often be active at the optimum. The problem is determining which of the inequality constraints is active at the optimum, and then this inequality becomes an equality condition on the solution.

Siddal (1975) recommends that a simple absolute value transformation be used on every variable, for example, $X_i = |X_i|$. If any of the variables slip into a negative-value realm, even temporarily, this can cause considerable computation problems. A mathematical justification for this practice indicates this would not change the end result.

As Siddal (1975) has stated: "One man's optimization criterion is often another man's constraint." Certainly, one of the more troublesome

209

problems to the beginning designer (and even to the experienced one) is determining which items should go into the objective function and which should appear in the constraint. In typical problems, some variables will be quite obvious in this regard and others will be totally obscure. It is possible that a constraint of seemingly small effect may unreasonably limit the solution of a problem. *Trade-off functions* (Fox, 1971) and *utility functions* (Siddal, 1974) are two approaches that may help in the definition of objective functions and constraints.

Confusion can arise in the determination of which variables can be considered *design variables,* as some of these parameters are always present in design problems (Siddal, 1975). The complication comes when *state variables* are present in a design problem. State variables are usually defined as parameters that are set by the state or condition of the component or subsystem. Examples include temperature, pressure, enthalpy, and many others. A rule of thumb is that if the variable can be directly affected by a design choice, such as the pressure being set by the appropriate choice of pump parameters, then that variable is a design variable. Often, state variables can be eliminated from the objective function and the constraints, but sometimes, their forms are too complex to accomplish this. When state variables cannot be eliminated, for whatever reason, they must be treated as design variables.

Attempts have been made to solve problems with multiple objective functions. For example, one may wish to minimize capital and operating costs as well as water use. Ito et al. (1982) have applied this type of approach for the design analysis of a piping system to minimize both the heat loss and the amount of insulation used. A common approach in these types of problems is to sum all objective functions, each with an appropriate weighting factor, and minimize the resulting composite function. This can lead to problems. Westerberg (1981) suggests it is best to solve each of the problems separately and use judgment in selecting the solution for the composite problem.

Empirical expressions, special approximations, charts, and tables can offer special challenges. Many empiricisms used by experienced designers may not be easily made quantitative. Approximations may hold very well in certain regions of the solution but may give unreasonable results in other areas. While the conversion of tabular or graphical data to an appropriate functional form has been addressed in an earlier chapter, be aware that some of this data may have actually been generated from a functional form originally. Some background searching may yield the original function and save a great deal of time.

A very frustrating problem can be the realization that a bad starting point has been used for the optimization solution. At least two techniques can be used to help get the right start. One of these is to start with an artificial objective function made from a weighted sum of the constraints of the original problem. Moving in the direction of the gradient of this specially constructed function will often yield a desirable start to the original problem. A second method for getting started is to convert the constraints into differential equations that are then solved for the approximate solution to the initial problem. For certain special conditions, this will yield the same result as the Newton-Raphson equations (Westerberg, 1981).

Never rely completely on the results as they come from the computer. The old adage "garbage in, garbage out" needs to be heeded. Always check that your results reflect an element of reason.

9.8 SUMMARY COMMENTS

Optimization techniques is a topic with a great deal of previous work and where new understanding is being realized daily. The surface has barely been scratched in what is summarized here.

The works noted in the References are quite extensive. Many include listings of computer codes that can be used in a very direct fashion to solve a variety of problems. Your own computing equipment or facilities undoubtedly include some capabilities of this sort. A number of summaries of existing software have been given (e.g., Lasdon, 1981).

If you are a designer, it is imperative that you incorporate into your design analysis process the vast array of now available tools. Many other designers are doing this!

REFERENCES

Andeen, B., and L. Glicksman, 1972, "Computer Optimization of Dry Cooling Tower Heat Exchangers," ASME Paper 72-WA/Pwr-8.

Avriel, M., and D. Wilde, 1967, "Optimal Condenser Design by Geometric Programming," I&EC Process Design and Development, 6, pp. 256-263.

Beightler, C., and D. Phillips, 1976, APPLIED GEOMETRIC PROGRAMMING, Wiley, New York.

Crozier, R., 1980, "Designing a 'Near Optimum' Cooling-Water System," Chemical Engineering, April 2, pp. 118-127.

Dantzig, G., 1963, LINEAR PROGRAMMING AND EXTENSIONS, Princeton University Press, Princeton, NJ.

Dantzig, G., M. Dempster, and M. Kallio, Eds., 1981, LARGE-SCALE LINEAR PROGRAMMING, VOLUME 1, IIASA Collaborative Proceedings Series, CP-81-51, IIASA Laxenburg, Austria.

Fletcher, R., 1981, PRACTICAL METHODS OF OPTIMIZATION, VOLUME 2, CONSTRAINED OPTIMIZATION, Wiley, New York.

Fletcher, R., and C. Reeves, 1964, "Function Minimization by Conjugate Gradients," Computer Journal, 7, pp.149-154.

Fox, R., 1971, OPTIMIZATION METHODS FOR ENGINEERING DESIGN, Addison-Wesley, Reading, Mass.

Garg, H., U. Rani, and R. Chandra 1981, "Optimization of Fin and Tube Parameters in a Flat-Plate Collector," Energy, 6, pp. 83-92.

Gill, P., W. Murray, and M. Wright, 1981, PRACTICAL OPTIMIZATION, Academic, New York.

Henley, E., and R. Williams, 1973, GRAPH THEORY IN MODERN ENGINEERING, Academic, New York, Chapter 7.

Hrymak, A., G. McRae, and A. Westerberg, 1985, "Combined Analysis and Optimization of Extended Heat Transfer Surfaces," Journal of Heat Transfer, 107, pp. 527-532.

Ito, K., S. Akagi, and M. Nishikawa, 1982, "A Multiobjective Optimization Approach to a Design Problem of Heat Insulation for Thermal Distribution Piping Network Systems," ASME Paper 82-DET-57.

Karamchetty, S., 1980, "An Optimization Technique for Cogeneration System Simulation," ASME Paper 80-Pet-26.

Kroger, D., 1983, "Design Optimisation of an Air-Oil Heat Exchanger," Chemical Engineering Science, 38, pp. 329-33.

Lasdon, L., 1981, "A Survey of Nonlinear Programming Algorithms and Software," in FOUNDATIONS OF COMPUTER-AIDED CHEMICAL PROCESS DESIGN, VOLUME 1 (R. Mah and W. Seider, Eds.), Engineering Foundation, New York, pp. 185-217.

Lootsma, F., 1973, "A Survey of Methods for Solving Constrained-Minimization Problems via Unconstrained Minimization," in OPTIMIZATION AND DESIGN (M. Avriel, M. Rijckaert, and D. Wilde, Eds), Prentice-Hall, Englewood Cliffs, NJ, pp. 88-118.

McConnell, J., and L. Elmenius, 1976, "A Systematic Approach to the Economic Selection of Design Parameters for an Industrial Power Plant," Combustion, 48, December, pp. 7-12.

Mott, J., J. Pearson, and W. Brock, 1972, "Computerized Design of a Minimum Cost Heat Exchanger," ASME Paper 72-HT-26.

Nishio, M., I. Koshijimma, K. Shiroko, and T. Umeda, 1985, "Synthesis of Optimal Heat and Power Supply Systems for Energy Conservation," I&EC Process Design and Development, 24, pp. 19-30.

Paul, H., 1982, "An Application of Geometric Programming to Heat Exchanger Design," Computers and Industrial Engineering, 6, pp. 103-114.

Peterson, E., 1973, "An Introduction to Mathematical Programming," in OPTIMIZATION AND DESIGN (M. Avriel, M. Rijckaert, and D. Wilde, Eds.), Prentice-Hall, Englewood Cliffs, NJ, pp. 7-36.

Reklaitis, G., A. Ravindran, and K. Ragsdell, 1983, ENGINEERING OPTIMIZATION, Wiley-Interscience, New York.

Ravindran, A., 1982, "Linear Programming," Chapter 14 in HANDBOOK OF INDUSTRIAL ENGINEERING (G. Salvendy, Ed.), Wiley, New York, pp. 14.2.1-14.2.11.

Ray, W., B. Jung, and W. Mirosh, 1971, "Large Scale Process Optimization Techniques," Canadian Journal of Chemical Engineering, 49, pp. 844-852.

Rikhter, L., E. Volkov, E. Gavrilov, V. Lebedev, and V. Prokhorov, 1976, "Determining the Cost of Thermal Power Station Chimneys and Optimising the Gas Velocities in the Flue," Combustion, 48, December, pp. 34-40.

Sargent, R., 1973, "Minimization without Constraints," in OPTIMIZATION AND DESIGN (M. Avriel, M. Rijckaert, and D. Wilde, Eds.), Prentice-Hall, Englewood Cliffs, NJ, pp. 37-75.

Sargent, R., 1980, "A Review of Optimization Methods for Nonlinear Problems," in COMPUTER APPLICATIONS TO CHEMICAL ENGINEERING, PROCESS DESIGN AND SIMULATION (R. Squires and G. Reklaitis, Eds.), American Chemical Society, Washington, DC, pp. 37-52.

Siddal, J., 1974, ANALYTICAL DECISION-MAKING IN ENGINEERING DESIGN, Prentice-Hall, Englewood Cliffs, NJ.

Siddal, J., 1975, "Some Practical Problems in Optimization," ASME Paper 75-DET-107.

Straeter, T., 1969, "A Comparison of Gradient Dependent Techniques for the Minimization of an Unconstrained Function of Several Variables," Paper 69-951, AIAA Aerospace Computer Systems Conference, September.

Umeda, T., and A. Ichikawa, 1971, I&EC Process Design and Development, 10, p. 229.

Vanderplaats, G., 1984, NUMERICAL OPTIMIZATION TECHNIQUES FOR ENGINEERING DESIGN WITH APPLICATIONS, McGraw-Hill, New York.

Vinante, C., and E. Valladares, 1985, "Application of the Method of Multipliers to the Optimization of Chemical Processes," Computers and Chemical Engineering, 9, pp 83-87.

Westerberg, A., 1981, "Optimization in Computer Aided Design," in FOUNDATIONS OF COMPUTER-AIDED CHEMICAL PROCESS DESIGN, VOLUME 1 (R. Mah and W. Seider, Eds.), Engineering Foundation, New York, pp. 149-183.

PROBLEMS

9.1 Minimize the function

$$z = (y - x^2)^2 + 0.01(x - 1.0)^2$$

from the starting point $x = -1.2$ and $y = 1.0$. Note that this function exhibits a long parabolic valley along $x = y^2$ (Straeter, 1969).

9.2 Plot (i.e., show contour lines of the objective function and the constraints) the following system.

Minimize $\quad\quad z = -x y e^{-(x+y)^2}$

Subject to $\quad\quad y = 0.2 + x$
$$x \geq 0$$
$$y \geq 0$$
$$y \leq 0.1 / x$$

Crosshatch the area that could contain the solution as the result of satisfying the constraints. Then denote the location of the optimum.

9.3 Using a numerical approach, find the solution to Problem 9.2.

9.4 Consider the function $f(x) = x^3 - 0.5 x^2 - 10 x + 4$ in the range $-8 > x > 8$. Find the following items by a numerical method and check the results analytically.

   (a) All local maxima.
   (b) All local minima.
   (c) The global maximum.
   (d) The global minimum.

9.5 Using a numerical method appropriate for constrained optimization, solve the following minimization problem.

$$\text{Minimize} \qquad y = \frac{1}{(x - 1)(x - 6)^2}$$

$$\text{Subject to} \qquad y \geq 0.01 + 0.005 x^2$$

9.6 Find the minimum of the function $z = 4x^2 + 3xy + 7y^2$ using a numerical technique and starting from the point $x = -3$ and $y = 4$.

9.7 Show the solution to the following optimization problem graphically on an x-y plot. Determine the numerical solution.

$$\text{Maximize} \qquad z = 3x + 2y$$

$$\text{Subject to} \qquad \begin{aligned} y &\leq 10 \\ 3y + 2x &\leq 3 \\ y + x &\geq 1 \\ 2y - x &\geq -10 \end{aligned}$$

9.8 Using a Golden Section Search, find the minimum of the function $(10 - x)^2$ in the range $5 < x < 15$.

9.9 Using the tabulated values and a polynomial estimation of the generating function, estimate the maximum of the function in the range $0 < x < 6$.

x	0	3.5	5
y	15.65	29.68	-6.599

9.10 Find the minimum of $y = 10 - 4x + x^2$ in the range $0 \leq x \leq 10$.

9.11 Determine the global maximum of $\sin x + 2 \sin 2x - \sin 3x + \sin 4x$ in the range $0 \leq x \leq 1$ using a numerical technique (not a polynomial curve fit). Substantiate your result analytically.

9.12 Using a polynomial curve fit to the function shown in Problem 9.11, find the global maximum for the range given there. Check this result with the analytically determined value.

9.13 Consider the following data pairs:

x	0.0	0.1	0.3	0.9
y	0.0	0.9834	2.5755	1.7420

(a) Using a quadratic polynomial curve fit, estimate the maximum value of the unknown function.
(b) Using a cubic polynomial curve fit, estimate the maximum value of the unknown function.

9.14 Reconsider the cogeneration example given in Section 9.6.1. Show the functional effects of changing the relative efficiency of the conversion process to electricity (i.e., the efficiency of the portion of the plant after the boiler). Show your results in a plot.

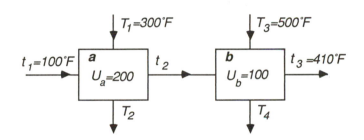

Figure 9.14 System used for solving Problem 9.15. The units on U are Btu/ft^2 hr °F.

9.15 In the system shown in Figure 9.14, determine the unknown temperatures so that the areas of heat exchangers **a** and **b** are minimized. Set up the minimization problem and solve a search numerically even though you may be able to solve the equations algebraically. The flow-rate-specific-heat product entering with each stream is given as follows. t_1 and t_2; mc_p= 2000 Btu/hr °F. T_1; mc_p= 1000 Btu/hr °F. T_3; mc_p= 1500 Btu/hr °F. The values for U in the figure have units of Btu/ft^2 hr °F.

9.16 In the system shown in Figure 9.15, determine the unknown temperatures so that the areas of heat exchangers **a, b, c,** and **d** are minimized. Set up the minimization problem and solve by means of a numerically based search. The flow-rate-specific-heat product entering with each stream is given as follows: t_1 and t_2; mc_p= 2000 Btu/hr °F. t_4

and t_5; mc_p= 2500 Btu/hr °F. T_1 and T_2; mc_p = 1000 Btu/hr °F. T_4 and T_5; mc_p= 1500 Btu/hr °F. The values for U in the figure have units of Btu/ft² hr °F.

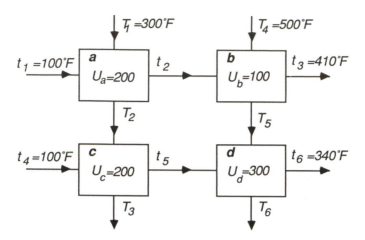

Figure 9.15 System to be used in solving Problem 9.16. After Henley and Williams (1973).

APPENDIX A

DESIGN PROJECT SUGGESTIONS

Note: Some possible project topics are listed below in no specific order. Virtually every project described is not fully defined. For each one of these, you are asked to perform a preliminary design. This means that approximations will have to be made, and assumptions will undoubtedly be needed. A key rule is: assume items only when necessary and then check the impact of the assumptions at some point in your analysis. Whereas some of the systems outlined below operate in a transient manner, all of these can be analyzed with a psuedo-steady-state model.

A.1 DESIGN OF A HEAT WHEEL REGENERATOR FOR HEAT RECOVERY

Some buildings have a requirement that a large amount of outside air must replace a fraction of the inside air each hour. In locations where ambient temperatures can vary considerably from desirable indoor temperatures, this requirement can be costly in terms of energy usage. The outside air must be appropriately conditioned before it is introduced inside. A "heat wheel" regenerator might be cost-effective. A device of this type consists of a wheel with spokes that rotate through an exhaust stream of a given flow rate and then through an outdoor ambient stream of the same flow rate. Heat is transferred to or from the spokes in one stream, and the opposite situation occurs in the other stream. Hence, the spokes should have both an effective heat transfer shape and a high heat capacity. For ease of analysis, assume that the spokes are solid cylinders of a metallic substance.

Determine design curves for least cost configurations as a function of the pertinent variable(s). For a large-capacity system, compare the cost of simply conditioning the ambient air to the cost of one of these designs. Assume that the air-handling system internal to the building is the same for both approaches. (This latter assumption may not be too accurate in some instances because the heat wheel configuration may require more ducting internal to the building.) If possible, compare your design to a commercial product.

A.2 HEAT-DRIVEN HEAT PUMP

Design a heat-driven heat pump that burns natural gas, and compare the costs of heating a building with this device in contrast to the direct use of gas. Consider two different gas heater efficiencies in your calculations: 50% (representative of many existing gas heaters) and 95% (more typical of many new designs). Are there significant cost and performance differences?

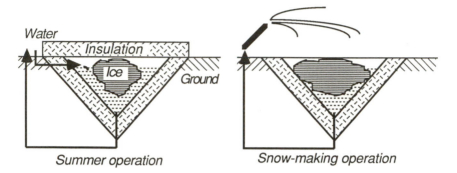

Summer operation   **Snow-making operation**

Figure A.1 Schematic of a storage system for the use of winter ice for summer air conditioning.

A.3 WINTER ICE FOR SUMMER AIR CONDITIONING

Some regions of the world have winters cold enough that ice can be formed by contacting water with the ambient air <u>and</u> summers warm enough to require a need for air conditioning. A scheme for taking advantage of the first of these characteristics to satisfy the need for the second has been proposed. The outdoor portion (see Figure A.1) of the system consists of an insulated storage tank located primarily below ground, a means of making ice (possibly a commercial snow-making machine), and a way of circulating the water, either to the air conditioning load or to the snow maker. You are to evaluate this type of approach for air conditioning for an application in Chicago. Assume that the storage container is conical with the apex pointed downward. An amount of insulation is applied in equal thicknesses on the sides and top of the storage. Using weather records for Chicago and (perhaps empirical) information about snow making, estimate the equivalent cost of this type of cooling. Treat the design of the storage (cone characteristics and insulation type and thickness) parametrically and indicate whether or not there is an optimum design for the configuration described. Compare the cost of cooling for your optimum (or "best" by some definition) design with the equivalent cost of vapor compression refrigeration air conditioning based upon electrical costs in your area. Finally, speculate on whether or not there is a storage configuration that is more desirable than the one described.

A.4 SOLAR-DRIVEN IRRIGATION SYSTEM

Vegetable growers from the Central Valley of California approach you about the possible feasibility of using solar-driven water pumps. Not only are they interested in the cost trade-offs with electrically driven pumps, but they are also facing a hint that their power requirements might not be met at some time in the future at any price. Configure a Rankine-type prime mover powered by the sun (see Figure A.2) and determine its cost as a function of the important parameter(s). Neglect any thermal storage in the system, but speculate what effects storage might have on the overall performance of the device. Assume that the pumped irrigation water can be used for cooling in the condenser. Compare this approach to the cost of electrically driven pumping over a 20-year life.

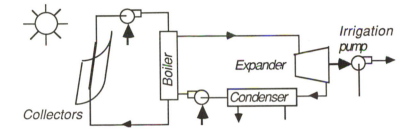

Figure A.2 A solar-driven irrigation system consists of an array of focusing collectors and no storage.

A.5 POWER GENERATION HEAT RECOVERY UNITS

A small company has approached you to perform a preliminary design of a unit to be installed on waste heat streams. The particular unit of interest is to reclaim heat at an average temperature of 250°F and generate electrical power. The company wishes to develop a product line around this concept but needs you to determine economic feasibility. Show the capacity effects on cost, and then try to estimate general trends in performance with source temperature level.

A.6 RESIDENTIAL ENERGY CONSERVATION OPTIONS EVALUATION

In your role as an engineer with a heating, ventilating, and air-conditioning firm, you are constantly being asked about the cost-effectiveness of various residential energy conservation options. Now is the time to evaluate key items. Compare the following items for new construction: various thicknesses of glass fiber insulation; various types of gas-fired furnaces; various efficiencies of air conditioners; and one-, two-, and three-paned windows, as well as movable insulation (assume this has an equivalent *R* value of 6 panes) over the windows.

A detailed analysis is not necessary. Instead, assume that your local climate can be represented by a fraction of the year at some winter design temperature, another fraction of the year at some summer design temperature, and the remainder of the year at a situation where there is no net heat transfer into or out of the building. The building can be assumed to be a square (in plan) structure of 2000 ft^2 of floor area, one story, with 8-ft ceilings and a flat roof, and windows that cover 20% of the wall areas. Assume there is no heat transfer through the floor. Neglect factors such as the precise makeup of the structure except for the amount of insulation present.

A.7 STEAM-INJECTED GAS TURBINE

It has been proposed that existing gas turbine engines might be modified to increase their output. Investigate the performance/cost feasibility of steam injection on a Brayton device. Is there an optimal design of such machines? Be careful to consider any physical limitations on the amount of water that might be injected. Try to fix some limits based upon physical factors of the machine. See Figure A.3 for a possible configuration of a steam-injected gas turbine. To simplify the analysis, assume that the turbine inlet temperature is always 2200°F. For cost considerations, assume that the basic Brayton cycle equipment has the same cost as stand-alone units but that the water system is additional.

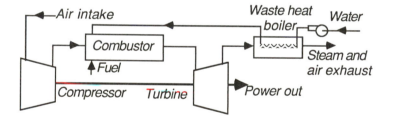

Figure A.3 A steam-injected gas turbine system.

A.8 PROPER AMOUNT OF PIPING INSULATION

As a part of the overall design of a new plant, the question naturally arose as to the proper amount of insulation to be used on the various pipe runs to be constructed. Both heated and cooled pipes are to be included, where flows from boilers and intermediate temperature sources are present as well as flows from chillers. Perform a design analysis of the optimal thickness of an appropriate pipe insulation of your choosing to be included on the pipes as a function of the pipe diameter and inner temperature.

Assume that the highest temperatures available in the plant are 1000°F, the coolest temperatures are 32°F, and all pipes are Schedule 80 construction with the largest size being 8 in. It can be further assumed that all pipes run through conditioned space at 80°F. **(a)** Assume that all devices are supplied by the boiler, which is fueled by natural gas (take the price as what your local supply costs), and absorption chillers furnish the cool temperature streams. **(b)** Would any of your results change if this exercise is performed for a power plant where a primary goal is the production of electrical power? Give reasons why or why not.

A.9 FLASHED GEOTHERMAL POWER PLANT

In many areas of the world, there are geothermal sources where hot water is available. There has been a great deal of work on the investigation of cost-effective approaches to the generation of power from these types of sources. One of the simpler type of approaches is the flashed-steam system. In this type of plant, the geothermal water comes from a well, it is flashed through a pressure drop (perhaps a valve), the liquid and vapor are separated (in a device like an insulated tank), and then the steam is circulated to a turbine. The spent steam is then condensed, and all of the liquid is generally reinjected into the ground at a significant distance from the source well.

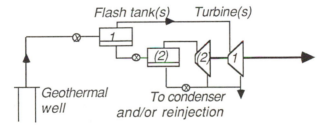

Figure A.4 A flashed-type geothermal power production plant is made up of a source of water, a flashing unit consisting of a pressure drop device and a separation tank, and a turbine-generator set. Shown here is a two-stage arrangement, but it can be a single-stage device. In the latter case, it would have only the first flash valve, flash tank, and turbine.

It has been proposed that a flashed-steam system be investigated for power production using geothermal water that is available from a vendor at the wellhead as slightly compressed liquid at 175°C. **(a)** Estimate the cost of steam for a single-flashed unit that would make power at the current average price in your region. **(b)** Next, repeat the same kind of estimate using a double-flashed unit. See Figure A.4. In both cases, show the effects of key parameters on the cost of water that is required, of course, including the output of the plant. Neglect any effects of corrosion and fouling, although in some situations, this could have a significant impact on costs.

Note that having the cost of the power generated from this device equal to the average price you pay may not make this approach currently competitive. If the geothermal source exists at this price, however, that kind of differential in cost might be competitive with other types of new generating plants. Also note that buying hot water or steam from a supplier who drills and maintains the wells is not an unusual situation. Many of the existing geothermal plants have this type of arrangement.

A.10 FLASH VERSUS BINARY GEOTHERMAL POWER PLANT

You are to compare the thermal efficiencies and total plant capital costs of both a binary geothermal plant and a single-flash geothermal plant for three sizes of output ratings: 0.1, 0.5, and 1 MWe. This is to be done for a range of geothermal source temperatures of 150, 175, and 200°C (all to be considered saturated liquid).

The flash unit can be considered similar to that shown in Figure A.4, but only with the components denoted as 1. The binary unit should be taken to be a simple Rankine system like that shown in Figure A.2; but instead of solar heated fluid on the input side of the primary heat exchanger, geothermal fluid will be used. Neglect any scaling and fouling aspects even though in some situations they can have considerable impact on the operations and maintenance of these kinds of plants. You might wish to try isobutane as the working fluid in the binary cycle.

A.11 DISTRICT HEATING/COOLING EVALUATION

With the development of cleaner conventional power plants and more interest in conservation of energy, the local electrical utility is considering building a new power plant and setting up a district heating operation in your local city (or town, as the case may be). The design of the primary plant will be very similar to other state-of-the-art facilities, but this one will have a turbine exhaust temperature of 250°F during times when there is heat demand in the city/town. The heat will then be piped to a grid for distribution to the users; and after use, the fluid is piped back to the plant. Does this type of scheme have promise? Make a preliminary design to see whether or not it does.

You may wish to treat the heat requirement by the users in a parametric manner. Variables that are investigated should include the use density (i.e., how much heat is required in a certain geographical area) and the manner the heat is transported (i.e., in steam form or by a liquid after heat exchange with the turbine exhaust steam). Be sure to include heat loss in your calculations.

Assume that the heat will not be of value if the carrier fluid drops below 150°F. Also assume that the plant will have a full heat rejection system whether or not the district heating system is installed.

A.12 HILLTOP TO VALLEY FLOOR POWER PLANT

An inventor approaches you with the observation that the temperatures at the mountaintops of certain western mountains are always considerably lower than the adjacent valleys. In these situations, the vertical difference in altitudes can be nearly 5000 ft for approximately the same amount of horizontal distance. The inventor knows that the laws of thermodynamics specify that there must be temperature difference present to be able to generate power. You are asked to evaluate this concept. A sketch of the system is as shown in Figure A.5. Take a preliminary evaluation of this concept. Assume that sufficient cooling (or "heating," as the case may be) water is available at each location at the local ambient temperature. (Although this may not actually be the case, the use of ambient air for cooling can only make the cost of such a plant higher. If the concept appears to be feasible, then the more realistic plant could be considered in another cut at a preliminary design.)

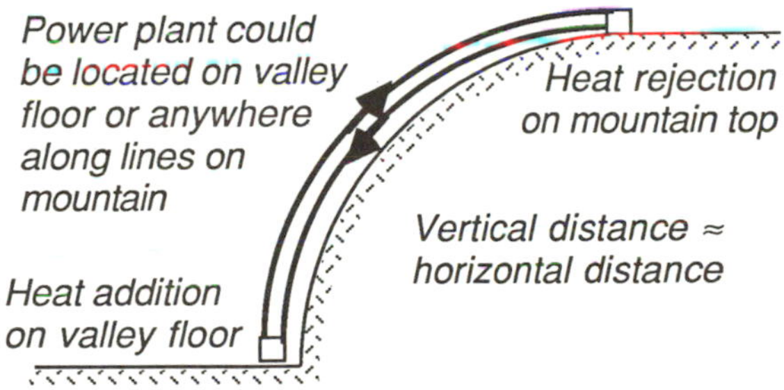

Power plant could be located on valley floor or anywhere along lines on mountain

Heat rejection on mountain top

Vertical distance ≈ horizontal distance

Heat addition on valley floor

Figure A.5 A sketch of a possible power plant that could be located somewhere between a heat rejection unit on a mountain top and a heat addition unit on an adjacent valley floor.

A.13 POWER PLANT DESIGN OPTIMIZATION

You are to select the optimal number of feedwater heaters and their extraction points for a Rankine cycle operating on coal. The cycle has a minimum pressure of 1 psia and a maximum pressure of 1000 psia. Temperatures to the primary turbine and exiting from the reheat loop are 1000°F. You should incorporate one heater of the open type, but this one will operate at atmospheric pressure, which is taken to be 14.7 psia. Assume that all turbines have an isentropic efficiency of 85% and all pumps are 87% efficient. A boiler efficiency of 90% can be taken here. Set the single stage of reheat such that the primary turbine exhaust is saturated vapor. Try to determine key component costs from Table D.3. Use the information on fuel costs from Chapter 6. The pricing of electrical power is a complex issue and should be simplified greatly for this project. Will a Second Law analysis lead to a different set of conclusions than will a First Law analysis?

A.14 STEAM-POWERED SPEED CAR

At the time of this writing, the speed record for a steam-powered car is 145.6 mph set on the Bonneville Salt Flats in a timed series of runs. You are to design a steam-propulsion system that has a good chance at breaking the record. You are completely free to make whatever design choices have to be made to do this, but costs cannot be unlimited. Specify the type of system to be used and size out the components. If weight is considered to be a factor, you may have to do more than a preliminary design to show the piping and other details. Be sure to compare the different types of approaches that have some possibility of success.

APPENDIX B
PROPERTY EVALUATION

B.1 COMPUTER EVALUATION OF PROPERTIES

The ability to evaluate properties within computer simulations is very important. In general, a variety of properties are needed. These include thermodynamic (pressure-temperature-density-entropy-enthalpy) properties, as well as thermal-physical information (including viscosity, thermal conductivity, and density). In what follows, property information for some liquids, vapors, and gases are given.

It is virtually impossible to give a listing of information that is comprehensive. No matter how many tabulations are given, at least one that is not included will be needed. Several references to information not listed here are given at the end of this appendix.

Property relationships that are needed and that are not presented here can be developed from tabulated data using curve fitting routines. This is discussed in Chapter 4. Many of the engineering texts contain data that can be used for developing computer routines via this method.

Accuracy of data and data fits will always be of concern. Comparing the values of properties presented in heat transfer texts, for example, will show that there can be considerable variation, unless all are taken from the same source. The development of an accurate data bank for property evaluation should be a high-priority goal.

B.2 CORRELATIONS

B.2.1 Thermal Conductivity

A listing of polynomial correlations for gas thermal conductivities has been given (Yaws, 1977). The correlating form is shown in Equation B.1.

$$k_G = a + bT + cT^2 + dT^3 \qquad (B.1)$$

Constants for a few of the substances tabulated by Yaws (1977) have been modified and are shown in Table B.1. Pressure does not affect the value of thermal conductivity over a wide range of values for many of the gases. For the constants shown in Table B.1, the units of k_G are W/m K and temperature should be in kelvin. See the original tabulation for detailed source and limitation information.

Table B.1
Thermal Conductivity Constants for Selected Gases and Vapors[a]
k_G, W/m K

Substance	$a \times 10^4$	$b \times 10^6$	$c \times 10^8$	$d \times 10^{12}$	298 K Value	T Range, K
Ammonia, NH_4	3.81	53.8	12.3	-36.3	0.02639	273↔1673
Carbon Dioxide, CO_2	-72.14	80.1	0.548	-10.5	0.0169	183↔1673
Helium, He	372	390	-7.49	12.9	0.147	113↔1073
Hydrogen, H_2	81.0	669	-41.6	156	0.175	113↔1473
Methane, CH_4	-18.7	87.3	11.8	-36.1	0.0337	273↔1273
Nitrogen, N_2	3.92	98.14	-5.07	15.0	0.0255	113↔1473
Oxygen, O_2	-3.27	99.7	-3.74	9.73	0.0263	113↔1473
Propane, C_3H_6	18.6	-4.70	21.8	-84.1	0.0176	273↔1273
Water, H_2O	73.4	-10.1	18.0	-91.0	0.0179	273↔1073

a. Data calculated from tabulation in Yaws (1977).

Similar correlations to those shown in Table B.1 have been given for liquids. Here a third-order polynomial suffices for the more limited temperature range. Hence, the thermal conductivities for the liquid state of the substances shown in Table B.2 are correlated with Equation B.2. Temperature is again used in units of kelvin.

$$k_L = a + bT + cT^2 \qquad (B.2)$$

Table B.2
Thermal Conductivity Constants for Selected Liquids[a]
k_L, W/m K

Substance	a	$b \times 10^2$	$c \times 10^4$	Reference Value		T Range,°C
Ammonia, NH_4	1.068	-0.158	-0.0123	0.501	@ 293 K	196↔373
Carbon Dioxide, CO_2	0.407	-0.0844	-0.0096	0.077	@ 293 K	217↔299
Helium, He	0.041	-1.832	37.89	0.0209	@ 3 K	2 ↔ 5
Hydrogen, H_2	-0.00855	1.036	-2.239	0.112	@ 23 K	14 ↔ 32
Methane, CH_4	0.3026	-0.0604	-0.03197	0.136	@ 153 K	90 ↔183
Nitrogen, N_2	0.263	-0.154	-0.00945	0.115	@ 90 K	64 ↔121
Oxygen, O_2	0.244	-0.0881	-0.0202	0.149	@ 90 K	55 ↔ 138
Propane, C_3H_6	0.261	-0.0531	-0.000888	0.098	@ 293 K	85 ↔ 193
Water, H_2O	-0.3838	0.5254	-0.0637	0.608	@ 293 K	273 ↔653

a. Data calculated from tabulation in Yaws (1977).

B.2.2 Heat Capacity

Heat capacity correlations for gases have been given by Yaws (1977). He used a third-order polynomial and found that average errors with this type of fit were typically much less than 1%. Use Equation B.3 with the data given in Table B.3. This equation is based upon T in kelvin.

$$c_{p,G} = a + bT + cT^2 + dT^3 \qquad (B.3)$$

Low pressures, typical of those associated with an ideal gas model, are assumed.

Table B.3
Heat Capacity Constants for Selected Gases and Vapors[a]
$c_{p,G}$, kJ/kg K

Substance	a	$b \times 10^3$	$c \times 10^6$	$d \times 10^9$	298 K
Ammonia, NH_4	1.49	2.02	-0.039	-0.162	2.09
Carbon Dioxide, CO_2	0.489	1.46	-0.945	0.230	0.848
Helium, He	5.20	0	0	0	5.20
Hydrogen, H_2	14.3	-0.0457	0.436	0.270	14.3
Methane, CH_4	1.32	2.43	2.32	- 1.40	2.23
Nitrogen, N_2	1.05	-0.196	0.493	-0.188	1.03
Oxygen, O_2	0.814	0.355	-0.048	-0.029	0.915
Propane, C_3H_6	-0.055	6.64	-3.12	0.621	1.66
Water, H_2O	1.88	-0.167	0.844	-0.270	1.90

a. Data calculated from tabulation in Yaws (1977). Temperature range of equation applicability is 298-1500 K. Pressures should be low so an ideal gas assumption applies.

For liquids, use Equation B.4 with T in kelvin. See Table B.4.

$$c_{p,L} = a + bT + cT^2 + dT^3 \qquad (B.4)$$

Table B.4
Heat Capacity for Selected Liquids, $c_{p,L}$, kJ/kg K[a]

Substance	a	$b \times 10^3$	$c \times 10^6$	$d \times 10^9$	$c_{p,L}$ Reference	T Range, K
Ammonia, NH_4	-8.05	130	-464.	576.2	4.40@-33.43°C	196↔373
Carbon Dioxide, CO_2	-80.82	1066	-4588	6588	1.93 @ -30°C	217↔293
Helium, He	-7.257	5804	-1227525	114×10^6	4.02@ -268.9	3↔ 6
Hydrogen, H_2	15.9	-1381	50967	-10196	8.8@-252.8°C	14↔ 28
Methane, CH_4	5.15	-43.26	301	-448	3.45@-161.5°C	90↔ 163
Nitrogen, N_2	-4.46	249	-3219	14060	2.05@-195.8°C	63 ↔ 113
Oxygen, O_2	-1.921	135.4	-1654	6598	1.70@-183°C	55 ↔143
Propane, C_3H_6	1.393	9.765	-55.95	126.3	2.23@-42.1°C	85↔353
Water, H_2O	2.823	11.83	-35.05	36.02	4.188@25°C	273↔623

a. Data calculated from tabulation in Yaws (1977).

B.2.3 Dynamic Viscosity

To find the dynamic viscosity of gases use Equation B.5 and Table B.5. The unit for T is kelvin.

$$\mu_G = a + bT + cT^2 \tag{B.5}$$

Table B.5
Dynamic Viscosity Constants for Selected Gases[a]
$$\mu_G, \text{ N·s/m}^2$$

Substance	a	$b \times 10^2$	$c \times 10^6$	μ_G @ 25°C	T Range, K
Ammonia, NH_4	-9.372	38.99	-44.05	103	73↔1473
Carbon Dioxide, CO_2	25.45	45.49	-86.49	153.4	173↔1673
Helium, He	54.16	50.14	-89.47	195.7	113↔1473
Hydrogen, H_2	21.87	22.2	-37.51	84.7	113↔1473
Methane, CH_4	15.96	34.39	-81.40	111.9	193↔1273
Nitrogen, N_2	30.43	49.89	-109.3	169.5	113↔1473
Oxygen, O_2	18.11	66.32	-187.9	199.2	113↔1273
Propane, C_3H_6	4.912	27.12	-38.06	82.4	193↔1273
Water, H_2O	-31.89	41.45	-8.272	90.14	273↔1273

a. Data calculated from tabulation in Yaws (1977).

Values for the dynamic viscosity of liquids are correlated with Equation B.6 and appropriate constants are given in Table B.6. As before, T is in kelvin.

$$\mu_L = a + b/T + cT + dT^2 \tag{B.6}$$

Table B.6
Dynamic Viscosity Constants for Selected Liquids[a]
$$\mu_L, \text{ N·s/m}^2$$

Substance	a	b	$c \times 10^2$	$d \times 10^6$	μ_L ref	T Range, K
Ammonia, NH_4	-8.591	876.4	2.681	-36.12	0.13@25°C	195↔141
Carbon Dioxide, CO_2	-1.345	21.22	1.034	-34.05	0.06@25°C	216↔242
Helium, He	-3.442	1.002	32.22	-35650		2↔5
Hydrogen, H_2	-4.857	25.13	14.09	-2773	0.016@-256°C	14↔33
Methane, CH_4	-11.67	499.3	8.125	-226.3	0.14@-170°C	90↔190
Nitrogen, N_2	-12.14	376.1	12.00	-470.9	0.18@-200°C	63↔77
Oxygen, O_2	-2.072	93.22	0.6031	-27.21	0.47@-210°C	55↔154
Propane, C_3H_6	-3.372	313.5	1.034	-20.26	0.091@25°C	84↔370
Water, H_2O	-10.73	1828	1.966	-14.66	0.90@25°C	273↔647

a. Data calculated from tabulation in Yaws (1977).

B.2.4 Latent Heat of Vaporization

A curve fit of data for latent heat of vaporization as a function of temperature has also been given by Yaws (1977). The form given by him is as in Equation B.7, and the correlation constants are in Table B.7. T can be in any set of units, but it, T_c, and T_1 must be in the same units. An alternative formulation for water is given later in this appendix.

$$h_{fg} = h_{fg,r} \left\lceil \frac{T_c - T}{T_c - T_1} \right\rceil^n \qquad (B.7)$$

Table B.7
Latent Heat of Vaporization Constants, h_{fg}, kJ/kg[a]

Substance	$h_{fg,r}$	T_1, K	T_c, K	n	T Range, °C
Ammonia, NH_4	1371	239.72	405.55	0.38	-77.4↔132.4
Carbon Dioxide, CO_2	235	273.15	304.25	0.38	-56↔26
Helium, He	21.8	4.25	5.15	0.38	-268.9↔-268
Hydrogen, H_2	44.8	20.35	32.9	0.237	-259.4↔-240.2
Methane, CH_4	510	111.65	190.55	0.38	-182.6↔-82.6
Nitrogen, N_2	199	77.35	126.35	0.38	-209.9↔-146.8
Oxygen, O_2	213	90.15	154.65	0.38	-218.4↔-118.5
Propane, C_3H_6	426	231.05	176.45	0.38	-187.7↔96.7
Water, H_2O	2256	373.15	647.35	0.38	0↔374.2

a. Data calculated from tabulation in Yaws (1977).

B.2.5 Density

Table B.8
Density Constants for Selected Liquids, ρ_L, kg/m³ [a]

Substance	a	b	T_c, K	Reference Value	T Range, °C
Ammonia, NH_4	231.2	0.2471	405.55	600 @ 298.15 K	-77.4↔132.4
Carbon Dioxide, CO_2	457.6	0.259	304.25	710 @ 298.15 K	-56.5↔31.1
Helium, He	74.7	0.4406	5.15	120@ 4.25 K	-271↔-268
Hydrogen, H_2	31.5	0.3473	32.9	70@ 20.37 K	-259.4↔-240.2
Methane, CH_4	161.1	0.2877	190.55	420@111.65 K	-182.6↔-82.6
Nitrogen, N_2	302	0.276	126.35	810@77.34 K	-209.9↔-146.8
Oxygen, O_2	422	0.2797	154.65	1140@90.0 K	-218.4↔-118.5
Propane, C_3H_6	220	0.2753	176.45	490 @ 298.15 K	-187.7↔96.7
Water, H_2O	347	0.274	647.35	1000 @ 298.15 K	0↔374.2

a. Data calculated from tabulation in Yaws (1977).

Values for densities of liquids can be correlated with a power law function:

$$\rho_L = A\, B^{-(1-T_r)^{2/7}} \qquad\qquad (B.8)$$

where T_r is the reduced temperature given by $T_r = T / T_c$. This form has been tabulated by Yaws (1977), and the appropriate constants are given in Table B.8. Absolute temperatures in any unit system can be used as long as they are the same as those used for the critical temperature.

Values of the densities of vapors can often be approximated by ideal gas behavior. If the thermodynamic state of the vapor is at least slightly removed from the saturation region, the errors involved will not be too great.

B.2.6   Other Properties

There are a number of sources of excellent information that can be used directly, or with some programming, to find properties. The largest body of information is available in the area of thermodynamic properties (Reynolds, 1979; Haar et al., 1984; Gonzalez-Pozo, 1986).

REFERENCES

Gonzalez-Pozo, V., 1986, "Formulas Estimate Properties for Dry, Saturated Steam," Chemical Engineering, May 12, p. 123.
Haar, L., J. Gallagher, and G. Kell, 1984, NBS/NRC STEAM TABLES, Hemisphere Publishing, Washington, DC.
Reynolds, W., 1979, THERMODYNAMIC PROPERTIES IN SI--GRAPHS, TABLES AND COMPUTATIONAL EQUATIONS FOR 40 SUBSTANCES, Department of Mechanical Engineering, Stanford University, Stanford, CA.
Yaws, C., 1977, PHYSICAL PROPERTIES, McGraw-Hill, New York, pp. 198-226.

APPENDIX C
CURVE FITTING CATALOG

C.1 INTRODUCTION

In what follows there are a number of correlations and descriptive plots of these correlations given. Use this information in matching distributions of physical data with appropriate curve fitting functions. Even if you have extensive data regression routines available, the information given here may be beneficial in attaining better "fits" if appropriate types of functions are chosen. (The term "fit" will be used as a substitute term for "curve fit.") The examples shown here are a limited selection of all possibilities, selected to cover only basic forms. Single-variable functions, $y = f(x)$, are shown.

C.2 CURVE FIT FORMS FOR SINGLE-VARIABLE FUNCTIONS

C.2.1 The Straight Line

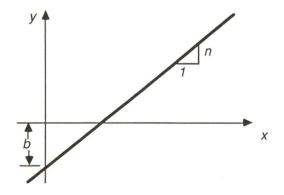

Figure C.1 A diagram of the fundamental linear form.

The fundamental function of engineering data is the straight line. A linear form occurs in so many situations it is included here simply to remind you to apply it whenever possible. A number of situations take place that, although nonlinear, may be fit quite well over some region of interest with a linear relationship. This fundamental form is shown in Equation C.1.

$$y = nx + b \qquad\qquad (C.1)$$

A graphical representation is found in Figure C.1. Of course, *n* is the *slope* of the line, and *b* is the *intercept*. Several of the equations that follow can be made linear by invoking certain transformations, and these will be noted.

C.2.2 Hyperbola Forms

The *hyperbola* is one typical case of nonlinear functions that can be linearized. Writing the general relationship for a hyperbola form, Equation C.2 results.

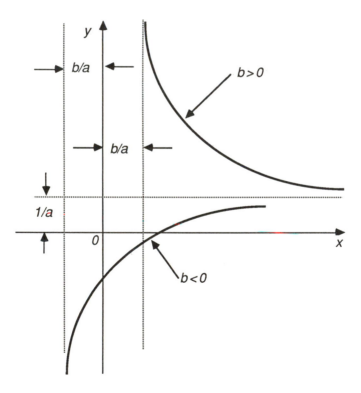

Figure C.2 Hyperbola forms of curves are shown, where *y* = *x/(ax - b)*. These curves are linear in the inverses of *y* and *x*. Hence, *(1/y) = c + d (1/x)*.

$$y = x / (ax-b) \tag{C.2}$$

To show a linear form in this relationship, take $1/y = u$ and $1/x = v$. An equation like that given in Equation C.1 will result. Examples of these characteristic curves are shown in Figure C.2.

C.2.3 Forms Involving Exponential Functions

The exponential function is used often to compress scales. A fundamental form of this function is shown in Equation C.3 and Figure C.3.

$$y = ae^{bx} \tag{C.3}$$

As was the case with hyperbola, the exponential functions are clearly nonlinear but can be made linear by an appropriate change in variable. A linear form results from taking the natural logarithm of both sides of the equation. Plotting functions of this form on semilog graph paper will result in a straight line.

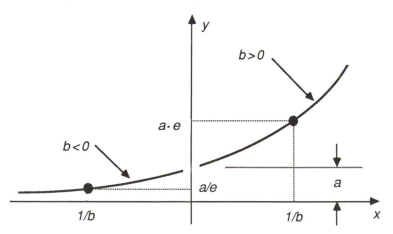

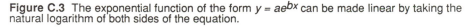

Figure C.3 The exponential function of the form $y = ae^{bx}$ can be made linear by taking the natural logarithm of both sides of the equation.

An equation that has important physical implications is the *exponential inverse.* This is shown in Figure C.4. Here the general function is given by Equation C.4.

$$y = ae^{b/x} \tag{C.4}$$

A related form representing characteristics that reach a limit asymptotically in a manner slightly different from Equation C.4 is given by Equation C.5, and this is shown in Figure C.5.

$$y = 1/ (a + be^{-x}) \tag{C.5}$$

233

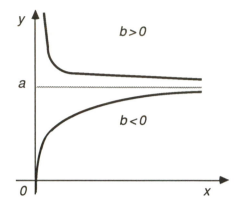

Figure C.4 Qualitative plots of two cases of the equation $y = ae^{b/x}$.

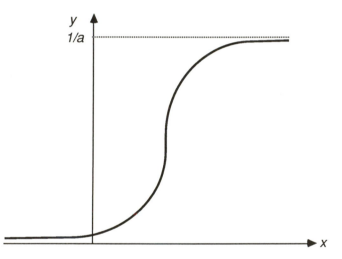
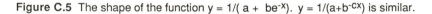

Figure C.5 The shape of the function $y = 1/(a + be^{-x})$. $y = 1/(a+b^{-cx})$ is similar.

C.2.4 Logarithmic Forms

Another nonlinear equation that can take on a linear form involves the log (or ln) functions. This includes all relationships of the form shown in Equation C.6. If *log x* is taken as the independent variable, rather than *x*, the equation is linear in this variable. See Figure C.6. Linear coefficients are found here also.

$$y = a + b \ (log \ x) \qquad (C.6)$$

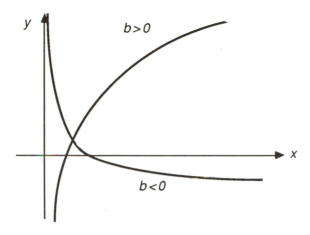

Figure C.6 Example forms of the equation $y = a + b \log x$.

C.2.5 Power Function Forms

A functional form that arises often in the correlation of heat transfer and fluid mechanics data is the power function, and a simple form of this function is shown in Equation C.7 and plotted in Figure C.7.

$$y = ax^b \tag{C.7}$$

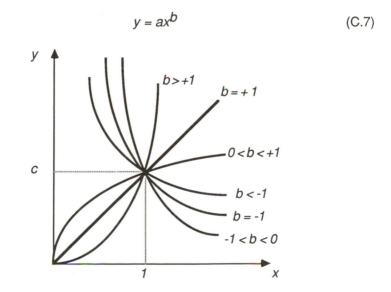

Figure C.7 The typical shapes of curves from power functions of the form $y = ax^b$, which can be linearized by taking the log (or ln) of both sides of the equation.

While y clearly varies in a nonlinear fashion with x, the equation can be cast into a linear form by taking the logarithm of both sides. The following equation is the result.

$$log\ y = log\ a + b\ log\ x \tag{C.8}$$

A related mathematical form to Equation C.5 is shown in Equation C.9. The qualitative curve for this equation is as shown in Figure C.5, although there are indeed differences in slopes between the functions shown in Equations C.9 and C.5.

$$y = 1/(a + b^{-cx}) \tag{C.9}$$

There is a very profound difference between Equations C.5 and C.9, however. Equation C.9 cannot be made linear in its coefficients, while Equation C.5 can be.

C.2.6 Polynomial Representations

Some of the most important forms for curve fitting are those available through a *polynomial representation.* A general form is given by Equation C.10.

$$y = \sum_{i=0}^{n} a_i x^i \tag{C.10}$$

From a practical standpoint, it is almost never the case that an order greater than four (i.e., $n > 4$) is used. In fact, most data will be fit by no greater than a second order form as shown in Equation C.11 and Figure C.8. Equation C.11 shows that while there is a nonlinear aspect between x and y that cannot be removed by a transformation, this equation is linear in its coefficients as shown.

$$y = a + bx + cx^2 \tag{C.11}$$

The case when the polynomial is made up of negative powers of x is also important in applications. This situation is indicated in Equation C.12.

$$y = a + b\,x^{-1} + c\,x^{-2} \tag{C.12}$$

This equation has a shape similar to the top portion of Figure C.4. As shown there, the asymptote at large values of x is a, and the function is undefined near the origin. If desired, a linear coordinate transformation of $x+d$ can be used for finite applications at the origin.

Other equations are possible. One that is used in applied mathematics to represent various functions is a *trigonometric series.* The power of this technique, known as the *Fourier series,* is widely known. It is linear in the coefficients, and an example can be written as

$$y = \sum c_i \sin^i x \tag{C.13}$$

Actually, the possiblities for combinations of successively higher powers of functions are very large. The application of combinations of functions in the fitting of data is a key technique used in many formal regression packages.

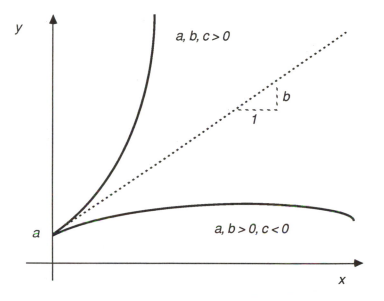

Figure C.8 A representation of the function $y = a + bx + cx^2$ in the first quadrant, showing the effect of the constants on the curve shape. This form has great utility in applications of curve fitting engineering data.

C.3 FUNCTIONS OF MORE THAN ONE VARIABLE

When more than one independent variable is involved, a virtual infinity of possibilities for fitting exists. These situations will not be dealt with here. Instead, treat these cases with the techniques described in Chapter 4.

C.4 FURTHER INFORMATION

For further information regarding curve fit forms and techniques beyond those discussed here and in Chapter 4, see the References of Chapter 4.

APPENDIX D
PRELIMINARY
COST ESTIMATION DATA

D.1 INTRODUCTION

In what follows, an extensive listing of cost function parameters is given. This was discussed in Chapter 6 and is listed in Table D.3. A key to section headings is found in Table D.1, and symbols are defined in Table D.2. Throughout Table D.3, there are footnotes given to references. Full listings of the references are given at the end of this appendix.

D.2 INDEX, DEFINITIONS, AND COST INFORMATION

Table D.1
Index to Cost Information Given in Table D.3

Listing Category	Page
Pumps, fans, blowers and compressors	239
Electrical motors, generators, and related	240
Electrical generation systems	240
Turbines and engines	240
Heat exchangers	241
Furnaces, boilers, heaters	242
Refrigerating systems, heat pumps	243
Miscellaneous	243
References	244

Table D.2
Definitions of Symbols in Table D.3

Symbol	Definition
C_r	Cost of reference size item, usually in multiples of $1000
m	Exponent on cost size relationship
$S\,Range$	Range of equipment sizes correlated
S_r	Reference size, in same units as $S\,Range$ values.
*	Additional option

238

Table D.3[a]
Equipment Cost Estimating Parameters

COMPONENT OR SYSTEM DESCRIPTION	m	C_r $1000	S_r	S Range, Units
PUMPS, FANS, BLOWERS AND COMPRESSORS				
PUMP, Centrifugal, horizontal, 50 ft head,	.26	2	10	.2↔16 kW
ci, radial flow, no motor. FOB. Others	.43	5.3	100	16↔400 kW
given in reference. {1}	.34	3.2	.5	.05↔30 m3/min
PUMP Same as above but w/motor. {1}	.39	2.5	10	1↔23 kW
	.58	7.5	100	23↔250 kW
	.59	4.3	1	.04↔30 m3/min
PUMP Centrifugal, axial flow, steel,				
w/motor, FOB. {1}	.03	47	10	4↔40 m3/min
PUMP Positive displacement, P_{in}=150 psi,				
P_{out}<1000 psi, no motor, gears. ci. FOB.				
(P_{out}=5000↔10000 psi, x2.5). {1}	.52	4	10	1↔70 kW
PUMP Diaphragm, large capacity, 15-ft				
head, w/motor. FOB. {1}	.46	4.3	.4	.06↔2 m3/min
FAN Centrifugal, radial bladed, 2.5 kPa. {2}				
Del., no motor.	.78	5.3	10	2↔100 nm3/s
Del., w/motor.	.93	9.3	10	2↔50 nm3/s
FAN Propeller w/ motor. FOB. {2}	.58	.45	1	.5↔6 nm3/s
	.36	1.5	10	6↔50 nm3/s
FAN See Vatavuk and Neveril (1981c); Means (1985).				
BLOWER Centrifugal, 28 kPa, del., no motor.	.61	160	30	12↔70 nm3/s
With motor, drive: 1.6 x (no motor cost). {2}				
BLOWER Rotary sliding vane, 275 kPa,				
del., no drive. {2}	.4	9.9	.1	.01↔.4 nm3/s
COMPRESSOR Centrifugal, <7000 kPa,				
del., w/ electric motor. {2}	.9	450	10^3	2↔4000 kW
COMPRESSOR Same as above, but no				
motor. FOB. {2}	.53	290	10^3	(5↔40)x10^2 kW
Exit pressure (MPa) factors: 1.7, x.8;				
6.9, x1; 14, x1.15; 34, x1.4; 48, x1.5.				
COMPRESSOR Completely installed				
compressor station, with land, either	.65	960	10^3	(1.5↔25)x10^2 kW
reciprocating or centrifugal, electric. {2}	.70	5900	10^4	(3↔13)x10^3 kW
COMPRESSOR Reciprocating <7000 kPa w/				
electric motor and aftercooler. FOB. {2}	.79	61	10^2	1↔10^3 kW
COMPRESSOR As above but no motor. {2}	1.0	21	10^2	10↔10^3 kW
Correction factors for the reciprocating listings:				
Installed=FOBx2. With steam turbine drive, x1.4.				
With gas engine drive, x2.8 Higher pressure				
output: 17 Mpa, x1.25 to 69 MPa, x1.95. Nearly				
linear variation in factor with pressure out.				
COMPRESSOR Air. {3}	.3	29	70	55↔1000 cfm
COMPRESSOR Air. See Means (1985) for detailed tabulations.				

a. Numbers in {} denote references given in footnotes at the end of each subsection. Abbreviations are defined in first set of footnotes. All costs adjusted for M&S index = 800.

Footnotes: {1} Woods et al.,1979a; {2} Woods et al., 1978; {3} Peters and Timmerhaus, 1980. Abbreviations used: ci=cast iron; cs=carbon steel; ss=stainless steel; w/=with; m3/min denotes cubic meters per minute of feed flow; del.=delivered; nm3/s is used in Woods et al. (1978) to denote normal (0°C and 760 mm of Hg) cubic meters per second. 1 nm3=37.32 scf; sat=saturated.

Table D.3 (Cont.)
Equipment Cost Estimating Parameters

COMPONENT OR SYSTEM DESCRIPTION	m	C_r $1000	S_r	S Range, Units
ELECTRICAL MOTORS, GENERATORS AND RELATED				
MOTORS AC, enclosed, fan cooled. {1}	.68	.67	10	$1 \leftrightarrow 10$ HP
(Other types listing in {1}.)	.87	.67	10	$10 \leftrightarrow 1000$ HP
GENERATORS AC electrical. {2}	0	1.9	-	$5 \leftrightarrow 10$ kW
(Another source is {3}.)	.66	1.9	10	$10 \leftrightarrow 1000$ kW
	.95	3.7	10^3	$10^3 \leftrightarrow 2.5 \times 10^4$ kW
GENERATORS Roesel. See Segaser (1977a).				

Footnotes: {1} Peters and Timmerhaus, 1980; {2} Dunn, 1979; {3} Segaser, 1978.

ELECTRICAL GENERATION SYSTEMS				
CONVENTIONAL POWER PLANTS $910/kW for 800 MW {1}. See also {2}.				
DRY STEAM TURBINE GENERATOR As				
used in Geysers geothermal field. {3}	.85	580	1	$10 \leftrightarrow 200$ MW
GEOTHERMAL PLANT See {3}, {4}, and {5}.				
NUCLEAR Complete system {2}				
$1750 \leftrightarrow 2550/kW, all sizes.				
CANDU heavy water, complete. {6}	.84	8.5E5	700 MW	$50 \leftrightarrow 100$ MW
Light water reactor, x.88 CANDU {6}				
SOLAR CENTRAL RECEIVER				
Total energy system. {7}	.62	3200	.3	$.3 \leftrightarrow 4$ MW
Other solar systems in {8}, {9}.				
COMBINED CYCLE Gas turbine/steam{10}	.93	3600	10	$3 \leftrightarrow 120$ MW
ENGINE/GENERATOR SETS 1200 rpm.				
Uninstalled w/o switch gear and controls. {11}	0	.25	1	$100 \leftrightarrow 600$ kW
FUEL CELL With waste heat recovery {7}	.8	18	25	$25 \leftrightarrow 250$ kW
ENVIRONMENTAL CONTROLS (See {12}, {13}.)				

Footnotes: {1} Anderson,1983; {2} Budwani, 1980; {3} Klei and Maslan, 1976; {4} Milora and Tester,1976; {5} Bloomster and Maeder, 1980; {6} Woods et al., 1979b; {7} Pine, 1979} {8} Barber, 1979; {9} Williams, 1980; {10} Christian, 1978b; {11}Segaser,1978b; {12} ASME,1981; {13} see all of the Vatavuk and Neveril listings.

TURBINES AND ENGINES				
STEAM TURBINE {1}	0	1.6	-	$5 \leftrightarrow 30$ kW
	.5	1.6	30	$30 \leftrightarrow 1000$ kW
(See also specific tabulations in {2}.)	.68	25	1000	$(1 \leftrightarrow 60) \times 10^3$ kW
Speed reducers. {3}	0	9	-	$5 \leftrightarrow 100$ kW
	.36	9	100	$100 \leftrightarrow 5000$ kW
GAS TURBINE Simple. {4}	.54	2300	15E3	$(1 \leftrightarrow 2) \times 10^4$ HP
	1.03	5600	4×10^4	$(2 \leftrightarrow 7) \times 10^4$ HP
GAS TURBINE Regenerative w/o generator.{4}	.57	4600	2×10^4	$(1.1 \leftrightarrow 4) \times 10^4$ HP
w/ generator unit. {4}	.53	6800	4×10^6	$(9 \leftrightarrow 80) \times 10^3$ kW
Speed reducers (based on plant kW). {3}	.55	7.3	100	$(1 \leftrightarrow 250) \times 10^2$ kW
DIESEL ENGINES Uninstalled, piston. {5}	.83	1200	5416	$(1 \leftrightarrow 15) \times 10^3$ HP
(O&M cost estimates tabulated also.)				
STEAM AND ORGANIC RECIPROCATING See {1}.				

Footnotes: {1} Mroz and Bailey, 1979; {2} Meador, 1978b; {3} Mroz,1979; {4} Farahan and Eudaly, 1978; {5} Segaser, 1977c.

240

Table D.3 (Cont.)
Equipment Cost Estimating Parameters

COMPONENT OR SYSTEM DESCRIPTION	m	C_f $1000	S_f	S Range, Units
HEAT EXCHANGERS (Costs can vary tremendously with material and flow design!)				
SHELL AND TUBE 150 psi.				
Floating head, cs, 16 ft long, del. {1}	.71	21	100	$2\leftrightarrow2000$ m^2
Factors: 400 psi, x1.25; 1000 psi, x1.55;				
3000 psi, x2.5; 5000 psi, x3.1.				
SHELL AND TUBE 304ss {2}	.67	.235	1	$400\leftrightarrow9000$ ft^2
See Woods, et al. (1976), Peters and Timmerhaus (1980), Corripio et al. (1982) and				
Purohit (1984) for detailed costing methods.				
PLATE AND FRAME CS				
frame, 304ss plates. {2}	.78	.1	1	$100\leftrightarrow5000$ ft^2
SPIRAL PLATE 304ss. {2}	.59	.66	1	$100\leftrightarrow1500$ ft^2
DOUBLE PIPE 600 psi cs. {1}	.14	1.1	2	$.3\leftrightarrow25$ m^2
AIR COOLER Finned tube, cs, 150 psi,				
includes motor and fan. FOB. {1}	.8	70	280	$20\leftrightarrow2000$ m^2
STEAM CONDENSER Water cooled. {3}	.55	3	10	$5\leftrightarrow10^5$ kW (plant)
STEAM CONDENSER Air cooled. {3}	.35	30	100	$5\leftrightarrow100$ kW (plant)
	.8	30	100	$10^2\leftrightarrow6\times10^5$ kW
BAROMETRIC CONDENSER Spray type, cs,				
w/o hotwell. {1}	.6	7.5	2	$.2\leftrightarrow40$ m^3/min
Same as above, but in terms of diameter.	1.4	12.3	1	$.1\leftrightarrow2.5$ m
HEAT RECOVERY UNIT For engine/generator.{4}	.45	.95	1	$200\leftrightarrow1500$ kW
Operating and maintenance costs=$.67/kWhr.				
HEAT RECOVERY UNIT Water and firetube				
boilers. {5} (Fluegas flow, scf/h)	.75	110	200	$30\leftrightarrow2000$ scf/hr
HEAT RECOVERY UNIT High-temperature				
fluid media on furnace. {6}	.68	3000	200	$30\leftrightarrow260$ MBtu/hr
AFTERCOOLER For compressor, cs,				
water on the shell side. {1}	.58	1	.1	$.03\leftrightarrow.3$ m^3/s
BAYONET HEATER For tank, FOB. {1}	.25	.21	.3	$.1\leftrightarrow1.3$ m^2
	1.33	1.9	5	$1.3\leftrightarrow12$ m^2
COIL IN TANK cs, FOB. {1}	-	.77		$.1\leftrightarrow30$ m^2
IMMERSION HEATER Electric, FOB. {1}	.87	1.9	50	$10\leftrightarrow200$ kW
COOLING TOWER Induced or forced draft, approach				
temp=5.5°C, wet-bulb temp.=23.8°C, range=5.5°C,				
directly installed, all costs except foundations,				
water pumps, and distribution pipes. {7}	1.	70	10	$4\leftrightarrow60$ m^3/min
	.64	560	100	$60\leftrightarrow700$ m^3/min
In terms of cooling capacity...................	1.	72	3.6×10^3	$10^3\leftrightarrow10^4$ kW

Factors to correct to other conditions:

Approach T, °C	Wet-bulb T, °C	Cooling range, °C
2.75, x1.50	10, x1.92	3, x0.78
4.0, x1.22	15, x1.43	5, x.92
7.0, x.85	20, x1.14	10, x1.3
11., x0.49	25, x.95	15, x1.62
14., x0.39	32, x.66	22, x1.93

	m	C_f $1000	S_f	S Range, Units
incl. foundations and basin x 1.7 to 3.0				
*Water distribution to and from				
cooling tower. Installed. {7}	.7	160	1	$.1\leftrightarrow2$ m^3/s
*Water treatment. Demineralizing, ion exchange,				
input 1330 ppm, output 30-40 ppm solids,				
installed. (FOB, x.7). {7}	1.	3200	.1	$.0004\leftrightarrow.8$ m^3/s
Factors to correct to other conditions Inlet feed:				
1000 ppm, x.5; 500ppm, x.25; 200ppm, x.18;100ppm, x.13.				
*Steam deaerator, cs,FOB. {7}	.78	67	(1)	$(.05\leftrightarrow40)\times10^5$ kg/hr
COOLING TOWER See also Vatavuk and Neveril (1981b).				

Footnotes: {1} Woods et al.,1976; {2} Kumana,1984; {3} Mroz and Bailey,1979; {4} Segaser, 1977b; {5} Vogel and Martin,1983; {6} Seifert et al.,1983; {7} Woods et al., 1979b.

Table D.3 (Cont.)
Equipment Cost Estimating Parameters

COMPONENT OR SYSTEM DESCRIPTION	m	C_f $1000	S_f	S Range, Units
FURNACES, BOILERS, HEATERS				
FLUIDIZED BED BOILER Burns high-sulfur				
coal, sat. steam, complete plant. {1}	.72	370	5000	$(5\leftrightarrow50)\times10^3$ lb/hr
Shop assembled unit, FOB. {1}	.6	150	5000	$(5\leftrightarrow50)\times10^3$ lb/hr
FIRETUBE PACKAGE BOILER FOB.{2}(See also {1}.)	.59	40	200	$40\leftrightarrow800$ HP
Stoker, economizer, dust collectors. {2}	.37	170	5000	$(2\leftrightarrow10)\times10^3$ lb/hr
	.56	500	25000	$(1\leftrightarrow5)\times10^4$ lb/hr
Gas fired, del., 50-200 psi sat. steam. {3}	.64	16	10^3	$(.2\leftrightarrow10)\times10^3$ kg/hr
WATER TUBE FOB. {2}	.67	340	12	$(4\leftrightarrow40)\times10^4$ lb/hr
WATER TUBE "D" TYPE FOB. {2}	.57	170	4×10^4	$(3\leftrightarrow7)\times10^4$ lb/hr
	.59	320	10^5	$(7\leftrightarrow18)\times10^4$ lb/hr
WATER HEATERS				
Gas-fired tank, FOB. {4}	1.1	.26	40	$30\leftrightarrow100$ gal
Electric heated tank, FOB. {4}	1.0	.26	50	$30\leftrightarrow100$ gal
Electric immersion, w/o tank, FOB. {5}	.87	1.3	50	$10\leftrightarrow200$ kW
Detailed listings in Means (1985). Large systems, see {1}.				
HEAT RECOVERY UNITS On furnace. {6}	.68	3000	200	$30\leftrightarrow260$ MBtu furnace duty
THERMAL-LIQUID HEATERS Gas fired. Linear variation of cost with duty between the two points: (1 MBtu/hr=$3700) and (10 MBtu/hr=$13,000). {1}				
ELECTRIC RESISTANCE HEATERS For household heating. Cost =$550 +$40/kW. {7}				
WASTE HEAT STEAM BOILER Unfired,				
150 psi, del. {3}	.81	160	10¢	$(.1\leftrightarrow10)\times10^4$ kg/hr
WOOD-FIRED BOILER {8}				
STEAM GENERATORS Pulverized coal stoker and oil fired. {9}				
BOILERS All types, see Means (1985).				
BOX-TYPE FURNACE 500 psi, cs, del. {5}	.75	144	12	$10\leftrightarrow400$ kW
Several other options costed in reference.				
BURNERS FOR FURNACE Includes blower				
and motor. Several items listed. {5}	.16	.53	.2	$.15\leftrightarrow.5$ kW
SOLID WASTE PYROLYZERS {10}				
COAL GASIFICATION {11}{15}				
INCINERATORS See also Means (1985).				
Rotary kiln, FOB. {12}	.54	1100	10	$4\leftrightarrow100$ MBtu/hr
Liquid injection, FOB. {12}	.45	220	4	$2\leftrightarrow100$ MBtu/hr
Solid waste w/ heat recovery. {13}				
FLARES Elevated, high Btu gas. {14}	.41	285	4E4	$(2.5\leftrightarrow250)$E5 lb/hr

Footnotes:{1} Blazek et al.,1979b;{2}Farahan, 1977a; {3} Woods et al.,1979b; {4} Farahan, 1977b; {5} Woods et al.,1976; {6} Seifert et al.,1983; {7} Goldman et al., 1977; {8} McGowan and Walsh,1980; {9} Coffin, 1984; {10} Boegly et al.,1977; {11} Blazek et al.,1979a) {12} Vogel and Martin,1983; {13} Boegly,1978; {14} Vatavuk and Neveril,1983b; {15} Baker, et al. 1979.

COMPONENT OR SYSTEM DESCRIPTION	m	C_f	S_f	S Range, Units
REFRIGERATING SYSTEMS, HEAT PUMPS				
AIR CONDITIONERS Room, FOB. {1}	.8	1.2	2	$.33\leftrightarrow15$ tons
Room, totally installed. {1}	.83	2.2	2	$.5\leftrightarrow15$ tons
Maintenance costs, $/year. {1}	.38	.2	2	$.33\leftrightarrow10$ tons
CENTRAL CHILLERS, VAPOR COMPRESSION {2}				
Reciprocating package, FOB.	.5	13.6	50	$10\leftrightarrow185$ tons
Roof reciprocating, air-cooled condensor, FOB.	.71	19.5	50	$20\leftrightarrow85$ tons
Centrifugal or screw compressor, FOB.	.66	92	500	$80\leftrightarrow2000$ tons
O+M annual costs, reciprocating.	.77	2	50	$10\leftrightarrow185$ tons
O+M annual costs, centrifugal or screw.	.42	8	500	$105\leftrightarrow2000$ tons

Table D.3 (Cont.)
COST ESTIMATING PARAMETERS

COMPONENT OR SYSTEM DESCRIPTION	m	C_r	S_r	S Range, Units
		\$1000		
REFRIGERATING SYSTEMS, HEAT PUMPS (Cont.)				
CENTRAL CHILLERS, VAPOR COMPRESSION {3}				
System installed, w/o cooling tower.	.77	267	1000	$20\leftrightarrow5000$ kW
CENTRAL CHILLERS, LiBr ABSORPTION {4}				
Single effect, equipment only.	.69	105	500	$100\leftrightarrow1400$ tons
Single effect, installed.	.66	160	500	$100\leftrightarrow1400$ tons
Double effect, equipment only.	.81	145	500	$400\leftrightarrow1200$ tons
Double effect, installed.	.7	230	500	$400\leftrightarrow1200$ tons
O+M, Single or double effect, yearly. {4}	.56	5.8	500	$100\leftrightarrow1400$ tons
LIQUID COOLER				
Intermittent surface conditioning system. {5}	.22	14	2×10^5	$(1.5\leftrightarrow20)\times10^5$ gal/d
COOLING TOWERS See listing under heat exchangers.				
SOLAR-DRIVEN COOLING MACHINES {6}				
Absorption machine, uninstalled.		25	25 tons	
Air-cooled Rankine driven vapor-compression.		38	25 tons	
AIR-TO-AIR HEAT PUMPS {7}				
Equipment only.	.86	2.4	3	$1\leftrightarrow50$ tons
Installed.	.9	4.9	3	$1\leftrightarrow50$ tons
O+M yearly costs.	.5	.3	3	$1\leftrightarrow50$ tons
WATER-TO-AIR HEAT PUMPS {8}				
Equipment only.	.64	1.65	3	$1\leftrightarrow25$ tons
Installed.	.69	3.4	3	$1\leftrightarrow25$ tons
Maintenance, years 2-5.	.5	.3	3	$1\leftrightarrow25$ tons
HEAT EXCHANGERS AND AIR CONDITIONERS Detailed listing in Means (1985).				

Footnotes: {1} Christian,1977c; {2} Christian, 1978a; {3} Woods et al.,1979b; {4} Christian,1977b; {5} Vatavuk and Neveril, 1983c; {6} Anand and Morehouse, 1980; {7} Christian,1977a; {8} Christian, 1977d.

MISCELLANEOUS				
STORAGE TANKS				
Vertical steel field erected tanks. {1}	.68	.017	1	$10^3\leftrightarrow10^5$ gal
Carbon steel. {2}	.56	1.4	100	$100\leftrightarrow10^5$ gal
304 ss or 30 psi cs (spherical). {2}	.52	2.6	100	$100\leftrightarrow10^5$ gal
Large volume cs, floating roof. {2}	.78	385	2×10^6	$(2\leftrightarrow10)\times10^6$ gal

Generally: {1}
Concrete \$.75-.90/gal
Fiberglass \$1.50/gal for 2000 gal size
Pipe type \$1.00/gal
More comprehensive method in Corripo et al. (1984).
PRESSURE VESSELS (Mulet et al.,1981)
PIPE INSULATION {3} (Contractor price)
    Elastomer 3/4-in. thickness. \$.52/ft;
    Phenolic foam, 1-in, thickness \$1.10/ft
    Fiberglass 1-in. thickness \$.70/ ft; Urethane \$1.00/ft
PIPE INSULATION Detailed listings in Means (1985).
    Rule of thumb "10% of total mechanical costs."
THERMAL STORAGE
    Low temperature. {1} High temperature. {4}
THERMAL CONVEYANCE SYSTEMS (Meador,1978a)
DUCTING (Vatavuk and Neveril,1980d; Means, 1985)
EXHAUST STACKS (Vatavuk and Neveril,1981d)
PIPING (Peters and Timmerhaus,1980; Culler,1984; Means,1985)
FUEL DISTRIBUTION AND STORAGE (Tison et al.,1979; Donakowski and Tison, 1979)
GAS PRODUCTION AND TREATMENT (Woods et al., 1982)

Footnotes: {1} Segaser and Christian,1979; {2} Peters and Timmerhaus,1980; {3} Boyd and Pesce, 1981; {4} Copeland et al., 1983.

REFERENCES

Anand, D., and J. Morehouse, 1980, "Economics of Solar Cooling," in SOLAR COOLING (Ed. by M. Nazer et al.), MRI/SOL-0601, pp.188-226.[1]

Anderson, R., 1983, "Estimate Capital, Operating Costs," Power, May, pp. 54-58.

ASME, 1981, "Estimated Costs of Environmental Control," Mechanical Engineering, December, p. 69.

Baker, N., C. Blazek, and R. Tison, 1979, "Low- and Medium-Btu Coal Gasification Processes," ANL/CES/TE 79-1, January.[1]

Barber, R., 1979, "Solar Rankine Engines--Examples and Projected Costs," ASME Paper 79-Sol-3.

Blazek, C., N. Baker, and R. Tison, 1979a, "High-Btu Coal Gasification Processes," ANL/CES/TE 79-2, January.[1]

Blazek, C., N. Baker, and R. Tison, 1979b, "Central Heating-Fossil-Fired Boilers," ANL/CES/TE 79-4, May.[1]

Bloomster, C., and P. Maeder, 1980, "Economic Considerations," in SOURCEBOOK ON THE PRODUCTION OF ELECTRICITY FROM GEOTHERMAL ENERGY (Ed. by J. Kestin et al.), DOE/RA-28320-2, pp. 682-713.[1]

Boegly, W., 1978, "Solid Waste Utilization-Incineration with Heat Recovery," ANL/CES/TE 78-3, April.[1]

Boegly, W., W. Mixon, C. Dean, and D. Lizdas, 1977, "Solid Waste Utilization-Pyrolysis," ANL/CES/TE 77-15, August.[1]

Boyd, L., and J. Pesce, 1981, "The Cost and Benefits of Solar Pipe Insulation," Solar Age, November, pp. 57-59.

Budwani, R., 1980, "Power Plant Capital Cost Analysis," Power Engineering, May, pp. 62-70.

Christian, J., 1977a, "Unitary Air-to-Air Heat Pumps," ANL/CES/TE 77-10, July.[1]

Christian, J., 1977b, "Central Cooling--Absorptive Chillers," ANL/CES/TE 77-8, August.[1]

Christian, J., 1977c, "Unitary and Room Air-Conditioners," ANL/CES/TE 77-5, September.[1]

Christian, J., 1977d, "Unitary Water-to-Air Heat Pumps," ANL/CES/TE 77-9, October.[1]

Christian, J., 1978a, "Central Cooling-Compressive Chillers," ANL/CES/TE 78-2, March.[1]

Christian, J., 1978b, "Gas-Steam Turbine Combined Cycle Power Plants," ANL/CES/TE 78-4, October.[1]

Coffin, B., 1984, "Compare Total Cost of Cogeneration-System Alternatives," Power, October, pp. 59-64.

Copeland, R., et al., 1983, THERMAL ENERGY STORAGE AT 900°C, SERI/TP-252- 2359.

Corripio, A., K. Chrien, and L. Evans, 1982, "Estimate Cost of Heat Exchangers and Storage Tanks via Correlations," Chemical Engineering, January 25, pp. 125-127.

Culler, D., 1984, "Pipe-Sizing Economics," Chemical Engineering, May 28, pp.

113-116.

Donakowski, T., and R. Tison, 1979, "Fuel Storage Systems," ANL/CES/TE 79-8, August.[1]

Dunn, J., 1979, "Commercial Synchronous Alternating-Current Generators," in HANDBOOK OF DATA ON SELECTED ENGINE COMPONENTS FOR SOLAR THERMAL APPLICATIONS, DOE/NASA/1060-78/1, pp. 175-186.[1]

Farahan, E., 1977a, "Central Heating--Package Boilers," ANL/CES/TE 77-6, May.[1]

Farahan, E., 1977b, "Residential Electric and Gas Water Heaters," ANL/CES/TE 77-2, August.[1]

Farahan, E., and J. Eudaly,1978, "Gas Turbines," ANL/CES/TE 78-8, October.[1]

Goldman, S., F. Best, and M. Golay, 1977, "End Use Space Conditioning Equipment Cost Data for Use in Total Energy System Analysis," AD-A042851, May.[1]

Klei, H., and F. Maslan, 1976, "Capital and Electric Production Costs for Geothermal Power Plants," Energy Sources, pp. 331-345.

Klepper, O., 1980, "Economics of Selected Fuels for Industrial Steam Capacity," ASME Paper 80-IPC/Pwr-5.

Kumana, J., 1984, "Cost Update on Specialty Heat Exchangers," Chemical Engineering, June 25, pp. 169-172.

McGowan, T., and J. Walsh, 1980, "An Economic Comparison of Wood and Fossil Fuel Processing System," ASME Paper 80-WAM.

Meador, J., 1978a, "Thermal Conveyance Systems," ANL/CES/TE 78-6, September.[1]

Meador, J., 1978b, "Steam Turbines," ANL/CES/TE 78-7, October.[1]

Means, 1985, MEANS MECHANICAL COST DATA, 1986, 9TH ANNUAL EDITION, R. S. Means Co., Kingston, MA.

Milora, S., and J. Tester, 1976, GEOTHERMAL ENERGY AS A SOURCE OF ELECTRICAL POWER, MIT Press, Cambridge, MA, Chapter 6 and Appendix D.

Mroz, T., 1979, "Speed Reducers-Increasers," in HANDBOOK OF DATA ON SELECTED ENGINE COMPONENTS FOR SOLAR THERMAL APPLICATIONS, DOE/NASA/1060-78/1, June, pp. 159-173.[1]

Mroz, T., and M. Bailey, 1979, "Rankine-Cycle Component Characteristics," in HANDBOOK OF DATA ON SELECTED ENGINE COMPONENTS FOR SOLAR THERMAL APPLICATIONS, DOE/NASA/1060-78/1, June, pp. 13-83.[1]

Mulet, A., A. Corripio, and L. Evans, 1981, "Estimate Costs of Pressure Vessels via Correlations," Chemical Engineering, October 5, pp. 145-150.

Peters, M., and K. Timmerhaus, 1980, PLANT DESIGN AND ECONOMICS FOR CHEMICAL ENGINEERS, THIRD EDITION, McGraw-Hill, New York.

Pine, G., 1979, "Economic Comparisons of Solar and Fossil Total Energy Systems for Industrial Applications," ASME Paper 79-WA/TS-6.

Purohit, G., 1984, "Estimating Costs of Shell-and-Tube Heat Exchangers," Chemical Engineering, August 22, pp. 56-67.

Segaser, C., 1977a, "Electric Generators--Roesel," ANL/CES/TE 77-3, April.[1]

Segaser, C., 1977b, "Heat Recovery Equipment for Engines," ANL/CES/TE 77-4, April.[1]

Segaser, C., 1977c, "Internal Combustion Piston Engines," ANL/CES/TE 77-1, July.[1]

Segaser, C., 1978, "Conventional Alternating-Current Generators and Engine Generator Sets," ANL/CES/TE 78-1, April.[1]

Segaser, C., and J. Christian, 1979, "Low Temperature Thermal-Energy Storage," ANL/CES/TE 79-3, March.[1]

Seifert, W., J. Beyrau, G. Bogel, and L. Wuelpern, 1983, "How to Evaluate Heat Recovery via High Temperature Fluid Media," Chemical Engineering, July 11, pp. 105-110.

Tison, R., N. Baker, and C. Blazek, 1979, "Fuel Distribution," ANL/CES/TE 79-7, July.[1]

Vatavuk, W., and R. Neveril, 1980a, "Estimating Costs of Air-Pollution Control Systems, I. Parameters for Sizing Systems," Chemical Engineering, October 6, pp. 165-168.

Vatavuk, W., and R. Neveril, 1980b,--, "II. Factors for Estimating Capital and Operating Costs," Chemical Engineering, November 3, pp. 157-162.

Vatavuk, W., and R. Neveril, 1980c,--, "III. Estimating the Size and Cost of Pollutant Capture Hoods," Chemical Engineering, December 1, pp. 111-115.

Vatavuk, W., and R. Neveril, 1980d,--, "IV. Estimating the Size and Cost of Ductwork," Chemical Engineering, December 29, pp. 71-73.

Vatavuk, W., and R. Neveril, 1981a,--, "V. Estimating the Size and Cost of Gas Conditioners," Chemical Engineering, January 26, pp. 127-132.

Vatavuk, W., and R. Neveril, 1981b,--, "VI. Estimating Cost of Dust-Removal and Water-Handling Equipment," Chemical Engineering, March 23, pp. 223-228.

Vatavuk, W., and R. Neveril, 1981c,--, "VII. Estimating Costs of Fans and Accessories," Chemical Engineering, May 18, pp. 171-177.

Vatavuk, W., and R. Neveril, 1981d,--, "VIII. Estimating Cost of Exhaust Stacks," Chemical Engineering, June 15, pp. 129-130.

Vatavuk, W., and R. Neveril, 1981e,--, "IX. Costs of Electrostatic Precipitators," Chemical Engineering, September 7, pp. 139-140.

Vatavuk, W., and R. Neveril, 1981f,--, "X. Estimating Size and Cost of Venturi Scrubbers," Chemical Engineering, November 3, pp. 93-96.

Vatavuk, W., and R. Neveril, 1982a,--, "XI. Estimate the Size and Cost of Baghouses," Chemical Engineering, March 22, pp. 153-158.

Vatavuk, W., and R. Neveril, 1982b,--, "XII. Estimate the Size and Cost of Incinerators," Chemical Engineering, July 12, pp. 129-132.

Vatavuk, W., and R. Neveril, 1982c,--, "XIII. Costs of Gas Absorbers," Chemical Engineering, October 4, pp. 135-136.

Vatavuk, W., and R. Neveril, 1983a, --, "XIV. Costs of Carbon Adsorbers," Chemical Engineering, January 24, p. 131.

Vatavuk, W., and R. Neveril, 1983b, --, "XV. Costs of Flares," Chemical Engineering, February 21, pp. 89-90.

Vatavuk, W., and R. Neveril, 1983c, --, "XVI. Refrigeration Systems,"

Chemical Engineering, May 16, pp. 95-98.

Vatavuk, W., and R. Neveril, 1984a, --, "XVII. Particle Emissions Control," *Chemical Engineering*, April 2, pp. 97-99.

Vatavuk, W., and R. Neveril, 1984b, --, "XVIII. Gaseous Emissions Control," *Chemical Engineering*, April 30, pp. 95-98.

Vogel, G., and E. Martin, 1983, "Estimating Capital Cost of Facility Components, Part 3," *Chemical Engineering*, November 28, pp. 87-90.

Williams, T., 1980, "Comparative Economics of Small Solar Thermal Electric Power Systems," 15th IECEC.

Woods, D., S. Anderson, and S. Norman, 1976, "Evaluation of Capital Cost Data: Heat Exchangers," *The Canadian Journal of Chemical Engineering*, 54, December, pp. 469-488.

Woods, D., S. Anderson, and S. Norman, 1978, "Evaluation of Capital Cost Data: Gas Moving Equipment," *The Canadian Journal of Chemical Engineering*, 56, August, pp. 413-435.

Woods, D., S. Anderson, and S. Norman, 1979a, "Evaluation of Capital Cost Data: Liquid Moving Equipment," *The Canadian Journal of Chemical Engineering*, 57, August, pp. 385-408.

Woods, D., S. Anderson, and S. Norman, 1979b, "Evaluation of Capital Cost Data: Offsite Utilities (Supply)," *The Canadian Journal of Chemical Engineering*, 57, October, pp. 533-565.

Woods, D., S. Anderson, and S. Norman-Sills, 1982, "Evaluation of Capital Cost Data: Onsite Utilities (Industrial Gases)," *The Canadian Journal of Chemical Engineering*, 60, April, pp. 173-201.

---

1. These documents are available from the National Technical Information Service, U. S. Deparment of Commerce, 5285 Port Royal Road, Springfield, Virginia 22161. Use the document number listed when ordering. Abstracts of several of these documents are given in Appendix E of this text.

APPENDIX E
PERTINENT GOVERNMENT PUBLICATIONS

Note: *There are literally tens of thousands of government documents, and a significant number of these could have some application to the design of thermal systems. While any short list of titles could not hope to be comprehensive, or even representative, some titles are given below. Generally, the reports listed have included design data or outline some potentially valuable design analysis tool(s). Unless otherwise indicated, these documents are available from National Technical Information Service, U.S. Department of Commerce, 5285 Port Royal Road, Springfield, Virginia 22161. Use the document number when ordering.*

Baker, N. R., Blazek, C. F., and Tison, R., "Coal Liquefaction Processes," ANL/CES/TE 79-6, July 1979. Coal liquefaction is still an emerging technology that is receiving great attention as a possible liquid fuel source. Currently, four general methods of converting coal to liquid fuel are under active development: (1) direct hydrogenation, (2) pyrolysis/hydrocarbonization, (3) solvent extraction, and (4) indirect liquefaction. This work is being conducted at the pilot plant stage, usually with a coal feed rate of several tons per day. Various processes are evaluated with respect to product compositions, thermal efficiency, environmental effects, operating and maintenance requirements, and cost.

Baker, N. R., Blazek, C. F., and Tison, R. R., "Low-and Medium-Btu Coal Gasification Processes," ANL/CES/TE 79-1, January 1979. Coal gasifiers, for the production of low- and medium-Btu fuel gases, come in a wide variety of designs and capacities. For single gasifier vessels, gas energy production rates range from about 1 to 18 billion Btu/day. The key characteristics of gasifiers that would be of importance for their application as an energy source in Integrated Community Energy Systems are evaluated here. The types of gasifiers considered here are single- and two-stage, fixed-bed units; fluidized-bed units; and entrained-bed units, as producers of both low-Btu (less than 200 Btu/SCF) and medium-Btu (200-400 Btu/SCF) gases. The gasifiers are discussed with respect to maximum and minimum capacity, the effect of feed coal parameters, product characteristics, thermal efficiency, environmental effects, operating and maintenance requirements, reliability, and cost.

Blazek, C. F., Baker, N. R., and Tison, R. R., "Central Heating--Fossil-Fired Boilers," ANL/CES/TE 79-4, May 1979. This evaluation provides performance and cost data for fossil-fuel-fired steam boilers, hot-water generators, and thermal fluid generators currently available from manufacturers. Advance technology fluidized-bed boilers also are covered. Performance characteristics that were investigated include unit efficiencies, turndown capacity, and pollution requirements. Costs are tabulated for equipment and installation of both field-erected and packaged units.

Blazek, C. F., Baker, N. R., and Tison, R. R., "High-Btu Coal Gasification Processes," ANL/CES/TE 79-2, January 1979. This evaluation provides estimates of performance and cost data for advanced technology, high-Btu, coal gasification facilities. The six processes discussed reflect the current state-of-the-art development. Because no large commercial gasification plants have yet been built in the United States, the information presented here is based only on pilot-plant experience. Performance characteristics that were investigated include unit efficiencies, product output, and pollution aspects. Total installed plant costs and operating costs are tabulated for the various processes.

Boegly, Jr., W. J., "Solid Waste Utilization-Incineration with Heat Recovery," ANL/CES/TE 78-3, April 1978. This evaluation considers the potential utilization of municipal solid wastes as an energy source by use of incineration with heat recovery. Subjects covered include costs, design data, inputs and outputs, and operational problems. Two generic types of heat recovery incinerators are evaluated. The first type, called a waterwall incinerator, is one in which heat is recovered directly from the furnance using water circulated through tubes imbedded in the furnace walls. This design normally is used for larger installations (>200 tons/day). The second type, a starved-air incinerator, is used mainly in smaller sizes (<l00 tons/day). Burning is performed in the incinerator, and heat recovery is obtained by the use of heat exchangers on the flue gases from the incinerator.

Boegly, W. J., Mixon, W. R., Dean, C., and Lizdas, D. J., "Solid Waste Utilization-Pyrolysis," ANL/CES/TE 77-15, August 1977. This evaluation considers the use of pyrolysis as a method of producing energy from municipal solid waste. Four processes are described in detail: "Purox" and "Refu-Cycler" produce a low Btu-gas; the "Occidental Process" produces an oil, and the "Landgard" Process produces steam using on-site auxiliary boilers to burn the fuel gases produced by the pyrolysis unit. Also included is a listing of other pyrolysis processes currently under development for which detailed information was not available. The evaluation provides information on the various process flowsheets, energy and material balances, product characteristics, and economics.

Christian, J. E., "Central Cooling - Absorptive Chillers," ANL/CES/TE 77-8, August l977. This technology evaluation covers commercially available single-effect Lithium-Bromide absorption chillers ranging in nominal cooling capacities of 3 to 1660 tons and double-effect Lithium-Bromide chillers from 385 to 1060 tons. Data are provided to estimate absorption chiller performance at off-nominal operating conditions. The part-load performance curves along with cost-estimating functions help the system design engineer select absorption equipment for a particular application based on life cycle costs. The Amnonia-Water absorption chillers are not considered to be readily available technology for ICES application; therefore, performance and cost data on them are not included in this evaluation.

Christian, J. E., "Central Cooling-Compressive Chillers," ANL/CES/TE 78-2, March 1978. The purpose of this report is to provide representative cost and performance data in a concise, usable form for three types of compressive liquid packaged chillers: reciprocating, centrifugal, and screw. The data are presented in graphical form as well as in empirical equations. Reciprocating chillers are available from 2.5 to 240 tons with full-load COPs ranging from 2.85 to 3.87. Centrifugal chillers are available from 80 to 2000 tons with full-load COPs ranging from 4.1 to 4.9. Field-assembled centrifugal chillers have been installed with capacities up to 10,000 tons. Screw-type chillers are available from l00 to 750 tons with full-load COPs ranging from 3.3 to 4.5.

Christian, J. E., "Gas-Steam Turbine Combined Cycle Power Plants," ANL/CES/TE 78-4, October 1978. The purpose of this technology evaluation is to provide performance and cost characteristics of the combined gas and steam turbine, cycle system applied to an Integrated Community Energy System (ICES). The basic gas steam turbine combined cycle consists of: (1) a gas turbine-generator set, (2) a waste heat recovery boiler in the gas turbine exhaust stream designed to produce steam, and (3) a steam turbine acting as a bottoming cycle. The total energy efficiency [(electric power = recovered waste heat) / (input fuel energy)] varies from about 65%-73% at full load to 34%-49% at 20% rated electric power output.

Christian, J. E., "Unitary Air-to-Air Heat Pumps," ANL/CES/TE 77-10, July 1977. This technology evaluation covers commercially available unitary heat pumps ranging from nominal capacities of 1-1/2 to 45 tons. The nominal COP of the heat pump models, selected as representative, vary from 2.4 to 2.9. Seasonal COPs for heat pump installations and single-family dwellings are reported to vary from 2.5 to 1.1, depending on climate. For cooling performance, the nominal EER's vary from 6.5 to 8.7. Representative part-load performance curves along with cost-estimating and reliability data are provided to aid: (1) the systems design engineer to select suitable-sized heat pumps based on life cycle cost analyses, and (2) the computer programmer to develop a simulation code for heat pumps operating in an Integrated Community Energy System.

Christian, J. E., "Unitary and Room Air-Conditioners," ANL/CES/TE 77-5, September 1977. The scope of this technology evaluation on room and unitary air conditioners covers the initial investment and performance characteristics needed for estimating the operating cost of air conditioners installed in an ICES community. Cooling capacities of commercially available room air conditioners range from 4000 Btu/h to 36,000 Btu/hr; unitary air conditioners cover a range from 6000 Btu/hr to 135,000 Btu/hr. The information presented is in a form useful to both the computer programmer in the construction of a computer simulation of the packaged air conditioner's performance and to the design engineer interested in selecting a suitably sized and designed packaged air conditioner.

Christian, J. E., "Unitary Water-to-Air Heat Pumps," ANL/CES/TE 77-9, October 1977. Performance and cost functions for nine unitary water-to-air heat pumps ranging in nominal size form 1/2 to 26 tons are presented in mathematical form for easy use in heat pump computer simulations. COPs at nominal water source temperature of 60°F range from 2.5 to 3.4 during the heating cycle; during the cooling cycle EERs range from 8.33 to 9.09 with 80°F entering water source temperatures. The COP and EER values do not include water source pumping power or any energy requirements associated with a central heat source and heat rejection equipment.

Cole, R. L., Nield, K. J., Rohde, R. R., and Wolosewicz, R. M., "Design and Installation Manual for Thermal Energy Storage," ANL-79-15, Second Edition, January 1980. The information contained in this manual includes sizing of storage, choosing of a location for the storage device, and insulation requirements. Both air-based and liquid-based systems are covered with topics included on designing rock beds, tank types, pump and fan selection, installation, costs, and operation and maintenance. Latent heat storage is also covered. The focus of the manual is on low to moderate temperatures as are often found in solar energy systems.

Donakowski, T. D., and Tison, R. R., "Fuel Storage Systems," ANL/CES/TE 79-8, August 1979. Storage technologies have been characterized for solid, liquid, and gaseous fuels. Emphasis is placed on storage methods applicable to systems based on coal. Items discussed here include standard practice materials and energy losses, environmental effects, operating requirements, maintenance and reliability, and cost considerations.

Farahan, E., "Central Heating - Package Boilers," ANL/CES/TE 77-6, May 1977. This report provides performance and cost data for electrical and fossil-fired package boilers currently available from manufacturers. Performance characteristics investigated include: unit efficiency, rated capacity, and average expected lifetime of units. Costs are tabulated for equipment and installation of various package boilers. The information supplied in this report will simplify the process of selecting package boilers required for industrial, commercial, and residential applications.

Farahan, E., "Residential Electric and Gas Water Heaters," ANL/CES/TE 77-2, August 1977. This report provides performance data for electric and gas-fired residential water heaters. Performance characteristics investigated include: unit full-load, part-load, and overall efficiencies and detailed examination of standby losses. Also included are brief discussions of energy-conserving options, such as lowering thermostat settings, increasing insulation thickness, and reducing pilot rate.

Farahan, E., and Eudaly, J. P., "Gas Turbines," ANL/CES/TE 78-8, October 1978. This evaluation provides performance and cost data for commercially available simple and regenerative cycle gas turbines. Intercooled, reheat, and compound cycles are discussed from theoretical basis only, because actual units are not currently available, except on a special-order basis. Performance characteristics investigated include unit efficiency at full-load and off-design conditions and at rated capacity. Costs are tabulated for both simple and regenerative cycle gas turbines. The output capacity of the gas turbines investigated ranges from 80 to 134,000 HP for simple units and from 12,000 to 50,000 HP for regenerative units.

Goldman, S. B., Best, F. R., and Golay, M. W., "End Use Space Conditioning Equipment Cost Data for Use in Total Energy System Analysis," FESA-RT-2038, May 1977. This report summarizes the capital cost data for end use space conditioning equipment used in possible Total Energy Systems. The capital costs are extracted from literature and vendor surveys, and an average "cost per unit of capacity" is derived. The end use equipment under study includes compressive air conditioners, electric baseboard heaters, heat pumps, and electric hot water heaters.

Kestin, J., et al., Eds., "Sourcebook on the Production of Electricity from Geothermal Energy," DOE/RA-28320-2, March 1980. Extensive sections are devoted to the following topics: resource characteristics; available work in geothermal energy; power systems (including thermodynamics, steam turbine characteristics, binary cycles and hybrid fossil-geothermal systems); waste heat rejection systems and equipment; materials selection; economic considerations; conceptual design optimization;

250

and environmental considerations.

Kreider, K. G., and McNeil, M. B., Eds., "Waste Heat Management Guidebook," NBS Handbook 121, February 1977. (*Purchase from the U.S. Government Printing Office, Washington, DC 20402.*) Sources of waste heat in industrial processes are reviewed, and an overview of off-the-shelf technology available for its use is given. Discussions of waste heat measurement technology and economics are included as are 14 case studies of successful industrial waste heat recovery installations.

Marshall, H. E., and Ruegg, R. T., "Energy Conservation in Buildings: An Economics Guildebook for Investment Decisions," NBS Handbook 132, May 1980. (*Purchase from the U.S. Government Printing Office, Washington, DC 20402.*) This guidebook provides principles, techniques, step-by-step illustrations, and sample problems on how to evaluate the economics of energy conservation and solar energy investments. Techniques of economic evaluation including life cycle costing, net benefits, savings-to-investment ratio, internal rate-of-return, and discounted payback analysis are described and compared in terms of their advantages and disadvantages. Discounting, a procedure for taking into account the time value of money, is illustrated in the analysis of an investment in heat pumps.

Meador, J. T., "Steam Turbines," ANL/CES/TE 78-7, October 1978. This report discusses the selection, classification, and average shaft efficiencies of some small-to-medium size steam turbines that can be used for base-load, turbine-generator units to meet both the electrical and thermal energy demands of several communities. Shaft efficiencies are evaluated by combining the average internal turbine efficiencies and the Rankine steam cycle efficiencies for several combinations of steam inlet conditions under conditions of both fully condensing and noncondensing exhaust steam. Efficiencies of very small turbines for mechanical drive of compressors or power plant auxiliaries also are estimated. A diagram based on vapor studies depicts the relationship of inlet steam pressure and degrees of superheat recommended for various exhaust temperatures. Some data on operational considerations and cost factors are included.

Meador, J. T., "Thermal Conveyance Systems," ANL/CES/TE 78-6, September 1978. This evaluation characterizes modern technology for: (1) long-distance, large-diameter (up to 5 ft), underground steam and high-temperature water (HTW) transport systems; and (2) hot-water and chilled-water systems that distribute thermal energy within communities. Because cost data on large sizes of commercially available pipe or prefabricated conduit are limited, this evaluation extrapolates available cost data and estimates the installed cost of some large, thermal conveyance systems. Data on cost in $10/Btu/mi for some steam and HTW conduit sizes are presented; both energy transport capability and relative cost of the two types of systems are compared.

Mroz, T., et al., "Handbook of Data on Selected Engine Components for Solar Thermal Applications," DOE/NASA/1060-78/1, NASA TM-79025, June 1979. This handbook provides a data base on developed and commercially available power-conversion-system components for Rankine and Brayton cycle engines, which have potential application to solar thermal power-generating systems. Design, performance, and cost data were provided by the respective manufacturers on steam turbines, reciprocating expansion engines, condensers, pumps, gas turbines, speed reducers, and AC generators. Components were selected for specific power levels from 5 to 50,000 kWe system output. Development status of the Stirling engine is included.

Reistad, G. M., and Means, P., "Heat Pumps for Geothermal Applications: Availability and Performance," DOE/ID/12020-T1, May 1980. The first section of this report considers the historical background, applications, achieved and projected performance evaluation, and performance improvement techniques. In the second section the commercial water source heat pump industry is considered in regard to both the present and projected availability and performance of units. In the final section, performance evaluations are made for units that use standard components but are redesigned for use in geothermal heating. Although all types of heat pumps are examined, the focus of the report is on unitary heat pumps for space heating applications. The scope of the report is quite broad and not restricted to geothermal applications.

Segaser, C. L., "Conventional Alternating-Current Generators and Engine Generator Sets," ANL/CES/TE 78-1, April 1978. The synchronous, rotating field alternator is the focus of this technology evaluation. Conventional 60-Hz, alternating-current generators, with standard ratings ranging from l.25 kVA to l0,000 kVA at voltages from l25 single-phase to l4,400 Volts three-phase and speeds up to l800 rpm, are covered. Technical data for representative diesel engine-generator sets for continuous prime power ratings up to 6445 kW are presented. Approximate l976 costs of standard electrical generating equipment are given for: (1) standard conventional alternating-current generators and (2) packaged engine-generator sets. The data indicate a decrease in unit costs as the power ratings increase, with the cost of the slow-speed units somewhat greater than that of the higher speed units. Maintenance data for a typical total energy plant presently in operation are given.

Segaser, C. L., "Electric Generators--Roesel," ANL/CES/TE 77-3, April 1977. A new and unique type of electrical generator is described that will provide constant-frequency output (within 0.01% or better) regardless of rotational speed variations. It accomplishes this with no added bulk over conventional generators, and with excellent efficiency. The diesel engine is a prime candidate for commercially available technologies for which part-load shaft efficiency is better at reduced speed than for constant-speed operation required by conventional AC generators. The unique characteristics of the Roesel generator allow prime mover speed to decrease with load and offer improved engine-generator system efficiency in ICES applications.

Segaser, C. L., "Heat Recovery Equipment for Engines," ANL/CES/TE 77-4, April 1977. The equipment usually employed to recover waste heat can be categorized as: (a) shell-and-tube-type heat exchangers, (b) radiator-type heat exchangers, (c) exhaust gas boilers for the generation of pressurized hot water and/or steam, (d) steam separators, and (e) combined packaged units for ebulliently cooled internal combustion piston engines. The functional requirements and cost considerations involved in applying these devices for the recovery of waste heat from various types of prime movers are examined in this evaluation.

Segaser, C. L., "Internal Combustion Piston Engines," ANL/CES/TE 77-1, July 1977. The key characteristics of internal combustion piston engines considered appropriate for use as prime movers in Integrated Community Energy Systems (ICES) are evaluated in this report. The categories of engines considered include spark ignition gas engines, compression ignition oil (diesel) engines, and dual-fuel engines. The engines are evaluated with respect to full-load and part-load performance characteristics, reliability, environmental concerns, estimated 1976 cost data, and current and future status of development. The largest internal combustion piston engines manufactured in the United States range up to 13,540 rated brake horsepower.

Segaser, C. L., and Christian, J. E., "Low Temperature Thermal-Energy Storage," ANL/CES/TE 79-3, March 1979. This report evaluates currently available techniques and estimated costs of low-temperature thermal energy storage (TES) devices applicable to Integrated Community Energy Systems (ICES) installations serving communities ranging in size from approximately 3000 (characterized by an electrical load requirement of 2 MWe) to about 100,000 population (characterized by an electrical load requirement of 100 MWe). Thermal energy in the form of either "hotness" or "coldness" can be stored in a variety of media as sensible heat by virtue of a change in temperature of the material, or as latent heat of fusion in which the material changes from the liquid phase to the solid phase at essentially a constant temperature.

Tison, R. R., Baker, N. R., and Blazek, C. F., "Fuel Distribution," ANL/CES/TE 79-7, July 1979. Distribution of fuel is considered here from a supply point to the secondary conversion sites and ultimate end users. All distribution is intracity with the maximum distance between the supply point and end use site generally considered to be 15 mi. Those fuels discussed are: coal or coal-like solids, methanol, No. 2 fuel oil, No. 6 fuel oil, high-Btu gas, medium-Btu gas, and low-Btu gas. Single-source, single-termination point and single-source, multitermination point systems for liquid, gaseous, and solid fuel distribution are considered. Transport modes and the fuels associated with each mode are: by truck -- coal, methanol, No. 2 fuel oil, and No. 6 fuel oil; by pipeline -- coal, methanol, No. 2 fuel oil, No. 6 oil, high Btu-gas, medium-Btu gas, and low-Btu gas. Data provided for each distribution system include component makeup and initial costs.

Yeoman, J. C., "Wind Turbines," ANL/CES/TE 78-9, December 1978. Wind turbines, ranging in size from 200 W to 10 MW, are discussed as candidates for prime movers in community systems. Estimates of performance characteristics and cost as a function of rated capacity and rated wind speed are presented. Data concerning material requirements, environmental effects, and operating procedures also are given and are represented empirically to aid computer simulation.

AUTHOR, SOURCE INDEX

SUBJECT INDEX